# Invariant Variational Principles

This is Volume 138 in
MATHEMATICS IN SCIENCE AND ENGINEERING
A Series of Monographs and Textbooks
Edited by RICHARD BELLMAN, *University of Southern California*

The complete listing of books in this series is available from the Publisher
upon request.

# INVARIANT VARIATIONAL PRINCIPLES

JOHN DAVID LOGAN

Department of Mathematics
Kansas State University
Manhattan, Kansas

ACADEMIC PRESS    New York   San Francisco   London    1977

A Subsidiary of Harcourt Brace Jovanovich, Publishers

ACADEMIC PRESS, INC.
111 Fifth Avenue, New York, New York 10003

*United Kingdom Edition published by*
ACADEMIC PRESS, INC. (LONDON) LTD.
24/28 Oval Road, London NW1


Library of Congress Cataloging in Publication Data

Logan, John David.
    Invariant variational principles.

    (Mathematics in science and engineering series ; v. 138)
    Bibliography:    p.
    Includes index.
    1.    Calculus of variations.    2.    Invariants.
3.    Transformations (Mathematics)    I.    Title.    II.    Series.
QA316.L76          515'.64          76-52727
ISBN 0−12−454750−8

To

Aaron and Rachel

# Contents

# Preface

This monograph deals almost exclusively with calculus of variations problems that are invariant under an $r$-parameter family of transformations. Since the early part of this century when Emmy Noether wrote her monumental paper on the subject, invariant variational principles have attained extensive application to a wide range of problems in physics and engineering; indeed, the Noether theorem has become one of the basic building blocks of modern field theories. The goal of this monograph is to transmit the flavor of some of these problems and applications and to present some concrete examples which it is hoped will lead the reader to appreciate the connection between invariance transformations and conservation laws for physical systems.

This monograph grew out of courses and seminars that the author conducted at the Ohio State University, the University of Dayton, the University of Arizona, and at Kansas State University over the past six years. Motivated by the belief that invariance problems do not get the attention they deserve in elementary courses in the calculus of variations, the author has injected much of this material in a standard course on the subject as a supplement to a basic textbook. The material has also been offered in a one-semester three-hour special topics course in applied mathematics. The audience has generally been upper division undergraduate students or graduate students in mathematics, physics, or applied science. The book is more or less self-contained and the prerequisites are minimal—undergraduate physics and standard courses in advanced calculus and linear algebra. There are several exercises at the end of each chapter ranging from routine to difficult, and some extend or generalize the results in the text.

The bibliography is by no means exhaustive. Rather, it has been selected to suit the needs of an introductory work. Many of the references cited are either standard or popular works, but they should provide the beginner with initial sources for more in-depth studies into the literature of the subject.

Every effort has been made to make the monograph as readable as possible for as wide an audience as possible. I have tried to avoid overly technical mathematical detail and notation in favor of a more classical style. However, resonable standards of rigor have not been sacrificed.

Chapter 1 is designed to review in detail the classical necessary conditions (the Euler–Lagrange equations) for extrema for both single and multiple integral problems in the calculus of variations and to briefly examine their role in physical theories by formulating Hamilton's principle for mechanical systems. It also establishes the notation, which I have tried to make as standard as possible, for the remainder of the book.

Chapter 2 discusses at length the famous Noether theorem for single integrals and concludes with a detailed application of the theorem by deriving the ten classical first integrals of the $n$-body problem from invariance under Galilean transformations. The approach to the Noether theorem presented here differs considerably from that in other texts. Our derivation is based on a set of fundamental invariance identities and has the advantage of being simpler and more direct than the usual argument based on "small variations" and the fundamental variational formula. These invariance identities not only lead to a direct proof of Noether's theorem, which in turn gives conservation laws, but they also have important consequences themselves; in Chapter 3 we show that these identities are similar to Killing's equations and that they lead to a group-theoretic method for obtaining first integrals for differential equations as well as a method for characterizing Lagrangians possessing given invariance properties. Chapter 4 repeats the results of Chapter 2 for the more-complicated multiple integral problem.

Chapter 5 contains an introduction to invariant variational principles in field theory. There are introductory sections on tensor algebra and the Lorentz group for those unfamiliar with these topics, and a brief review of electrodynamics is presented. Here, the notion of Lorentz invariance plays a decisive role; just as the Galilean transformations give rise to conservation laws for mechanics, so it is that the Lorentz transformations give rise to conservation laws for physical fields.

Chapter 6 introduces the idea of invariance under conformal transformations. Some introductory material is included on the form of these transformation in $R^n$ for definite and indefinite metrics. In Chapter 7 the results of earlier chapters are extended to a variational problem whose Lagrangian contains second-order derivatives; as an example, conservation laws for the Korteweg–deVries equation are obtained in this setting.

Chapter 8 deals with variational problems that are invariant under *arbitrary* transformations of the independent variables, or the so-called parameter-invariant problems. It is shown how the classical necessary conditions for parameter invariance, the Zermelo conditions, follow from fundamental invariance identities obtained in Chapter 2. Next, a rather heuristic discussion of the second Noether theorem, which involves transformations depending upon arbitrary functions, is presented. The discussion here, like that for the parameter-invariant problems, is quite abbreviated and is included only to introduce the topic.

# Acknowledgments

It pleases me to be able to thank some of my teachers, students, and colleagues whose influence appears in no small way in this little monograph. My own interest in this subject was stimulated by the lectures of Professors Stefan Drobot and Hanno Rund. My colleague, Professor W. D. Curtis at Kansas State University, and one of my students, Professor John S. Blakeslee, now at Washington State University, deserve particular thanks and credit for constantly prodding me in innumerable conversations to clarify my position on many matters in our joint effort to understand some of the fine points of the subject. Another colleague, Professor Chen-Jung Hsu, has also been a source of information on several occasions, and Professor Carl Rosenkilde read the manuscript and offered many helpful suggestions. I would like to thank Professor Richard Bellman for his continued interest in this project and, in particular, for his initial suggestion that I write such a work.

Finally, my thanks go to the typists, Ms. Judy Toburen and my wife, Marlyn, for their excellent work in typing the draft and the final manuscript.

CHAPTER **1**

# Necessary Conditions for an Extremum

## 1.1 INTRODUCTION

Quite generally, the calculus of variations deals with the problem of determining the extreme values (maxima or minima) of certain variable quantities called functionals. By a functional, we mean a rule which associates a real number to each function in some given class of so-called admissible functions. More precisely, let $A$ be a set of functions $x, y, \ldots$; then a functional $J$ defined on $A$ is a mapping $J: A \to R^1$ which associates to each $x \in A$ a real number $J(x)$. In the most general case, the set of admissible functions may, in fact, be any set of admissible objects, for example, vectors, tensors, or other geometric objects. A fundamental problem of the calculus of variations may then be stated as follows: given a functional $J$ and a well-defined set of admissible objects $A$, determine which objects in $A$ afford a minimum value to $J$. Here, we may interpret the word minimum in either the *absolute* sense, i.e., a minimum in the whole set $A$, or in the *local* sense, i.e., a relative minimum if $A$ is equipped with a device to measure closeness of its objects. For the most part, the classical calculus of variations restricts itself to functionals defined by integrals, either single integrals or multiple integrals, and to the determination of both *necessary* and *sufficient* conditions for a functional to be extremal.

1

Our primary interest in this monograph is what we shall term *invariance problems* in the calculus of variations. More precisely, we propose to investigate the consequences of the functional $J$ being invariant, in one sense or another, under special types of transformations. When studied in the context of a physical system, such invariance conditions have far-reaching consequences relative to the nature of the system itself. For example, we shall see that there is a strong connection between symmetries and conservation laws for a given system. And the application of invariance principles leads to identities which provide important information about both the variational problem defined by $J$ and its manifestation in a physical context.

Although our main interest is in invariance problems, it is nevertheless desirable to begin with a brief review of the classical necessary conditions for the basic variational problems that we shall study. This review will not only serve to set the stage for the investigation of invariance conditions, but it will also serve to establish terminology and notation, as well as make our work more or less self-contained. The reader desiring a more detailed account of the motivation and derivation of the classical necessary conditions for extrema is referred to one of the many treatises on the calculus of variations listed in the bibliography.

## 1.2  VARIATION OF FUNCTIONALS

We shall be somewhat general in our initial formulation of necessary conditions for extrema, saving particular functionals for later consideration. Therefore, we let a given, but arbitrary, functional

$$J:A \to R^1$$

be defined on some prescribed class of admissible objects $A$. We tacitly assume that $A \subseteq N$, where $N$ is a normed linear space; i.e., $N$ is a real linear space equipped with a mapping $\| \cdot \|: N \to R^1$ which associates to each $x \in N$ a nonnegative real number $\|x\|$, called the norm of $x$, satisfying the conditions

(1)   $\|x\| = 0$ if and only if $x = 0$;
(2)   $\|\alpha x\| = |\alpha| \|x\|$ for all $x \in N$ and all $\alpha \in R^1$;
(3)   $\|x + y\| \leq \|x\| + \|y\|$ for all $x, y \in N$.

Once $N$ is equipped with a norm, then it is possible to define the distance between two objects $x$ and $y$ in $N$; this distance is defined to be $\|x - y\|$. The motivation for assuming that the admissible set is contained in a normed linear space is that $A$ can then inherit these geometrical, or topological, properties of $N$, and consequently the notion of closeness between objects in $A$ as well as the consideration of continuous functionals on $A$ are feasible.

In most variation problems, the admissible set $A$ itself is usually not a linear space and the functionals $J$ are seldom linear functionals.

Now, an element $x \in A$ is said to be a *relative minimum* (with respect to the norm in $A$) of the functional $J$ if $J(x) \le J(y)$ for all $y \in A$ with $\| y - x \| < \delta$ for some $\delta > 0$. We seek a necessary condition for $x \in A$ to be a relative minimum. As might be suspected, the classical approach followed in the calculus for determining extrema of functions of real variables suggests a method for obtaining necessary conditions for functionals. In particular, we shall define a derivative of $J$ at $x$ and show that when $x$ is a relative minimum, then that derivative must vanish at $x$. For convenience, we denote the variational problem of determining the relative minima by

$$J(x) \to \min \qquad \text{on } A.$$

**1.1 Definition** Let $x_\varepsilon$ be a one-parameter family of elements in $A \subseteq N$, with $\| x_\varepsilon - x \| < \sigma$ for $|\varepsilon| < \varepsilon_0(\delta)$, of the form

$$x_\varepsilon = x + \varepsilon\eta,$$

where $x \in A$, $\eta \in N$, and $\sigma, \varepsilon_0(\delta) > 0$. Then the *first variation* of $J$ at $x$ in the direction $\eta$ is defined by

$$\delta J(x, \eta) = \left. \frac{dJ(x + \varepsilon\eta)}{d\varepsilon} \right|_{\varepsilon = 0} \tag{1.1}$$

if the derivative exists.

The derivative defined by (1.1) is commonly called a weak, or Gâteaux, derivative. When (1.1) is written differently as

$$\delta J(x, \eta) = \lim_{\varepsilon \to 0} \frac{J(x + \varepsilon\eta) - J(x)}{\varepsilon},$$

it shows itself as a directional derivative in the direction of $\eta \in N$. The assumption that $x$ is embedded in the one-parameter family $x_\varepsilon = x + \varepsilon\eta$ guarantees that there are enough "points" near $x$ in order to introduce a limiting process, i.e., a derivative at $x$. After Lagrange, we denote

$$\delta x \equiv \varepsilon\eta$$

and call $\delta x$ the variation of $x$.

As an immediate consequence of Definition 1.1 we have the following *necessary condition* for $x \in A$ to be a relative minimum.

**1.1 Theorem** If $x \in A$ is a relative minimum for the functional $J: A \to R^1$, then

$$\delta J(x, \eta) = 0 \tag{1.2}$$

for all $\eta \in N$ satisfying the assumptions of Definition 1.1. $\quad\square$

One might well ask if there exists a family $x_\varepsilon$ of the required form satisfying the conditions of this theorem. Fortunately, for the problems considered in the calculus of variations there are many $\eta$ for which (1.2) holds—so many, in fact, that $\eta$ will be able to be eliminated from the necessary condition. This remark manifests itself in the statement of the Fundamental Lemma of the calculus of variations, which will be proved in the next section.

## 1.3  SINGLE INTEGRAL PROBLEMS

We now specialize the results of the preceding section to functionals given by single integrals and defined on some space of functions. We begin with some definitions of the function spaces that will play a role in the analysis.

Let $C^2[a, b]$ denote the set of all continuous functions on the interval $[a, b]$ whose second derivatives are continuous, and let $C_n^2[a, b]$ denote the set of all vector functions

$$x(t) = (x^1(t), \ldots, x^n(t)), \qquad t \in [a, b],$$

whose components $x^k(t)$ belong to $C^2[a, b]$. Then $C_n^2[a, b]$ is a linear space with the obvious definitions of addition and scalar multiplication. Moreover, in $C_n^2[a, b]$ we can define a norm by

$$\|x(t)\| = \max_{t \in [a, b]} \{|x^1(t)|, \ldots, |x^n(t)|\} + \max_{t \in [a, b]} \{|\dot{x}^1(t)|, \ldots, |\dot{x}^n(t)|\}, \quad (1.3)$$

where $\dot{x}^k(t) = dx^k/dt$ denotes the derivative of $x^k(t)$. This norm, which is often called the *weak norm*, gives $C_n^2[a, b]$ the structure of a normed linear space. Two vector functions in $C_n^2[a, b]$ are close if both their components and the derivatives of their components are close.

Next, we define an admissible set of functions $A_n^2[a, b]$ by

$$A_n^2[a, b] = \{x(t) \in C_n^2[a, b] : x(a) = \alpha, x(b) = \beta, \alpha, \beta \in R^n\}. \quad (1.4)$$

Geometrically, the elements of $A_n^2[a, b]$ are the set of all smooth curves in $R^{n+1}$ which join the two fixed points $(a, \alpha)$ and $(b, \beta)$. If $\alpha = (\alpha^1, \ldots, \alpha^n)$ and $\beta = (\beta^1, \ldots, \beta^n)$, then the boundary conditions expressed in (1.4) are that the components of the admissible functions should satisfy

$$x^k(a) = \alpha^k, \qquad x^k(b) = \beta^k \qquad (k = 1, \ldots, n). \quad (1.5)$$

For conciseness, we shall in the sequel write $C_n^2$ and $A_n^2$ for $C_n^2[a, b]$ and $A_n^2[a, b]$, respectively.

Furthermore, let $L: R^1 \times R^n \times R^n \to R^1$ be a given function, called the *Lagrangian*, which is twice continuously differentiable in each of its $2n + 1$

arguments. For future reference, let us denote the arguments of the Lagrangian by $L = L(t, x^1, \ldots, x^n; y^1, \ldots, y^n)$. We then consider a functional $J: A_n^2 \to R^1$, called the *fundamental* or *action integral* defined by

$$J(x(t)) = \int_a^b L(t, x^1(t), \ldots, x^n(t); \dot{x}^1(t), \ldots, \dot{x}^n(t))\, dt$$

or, in vector notation,

$$J(x(t)) = \int_a^b L(t, x(t), \dot{x}(t))\, dt. \tag{1.6}$$

We are then faced with the variation problem of determining relative minima of $J$ in $A_n^2$, i.e.,

$$J(x(t)) \to \min \qquad \text{on } A_n^2.$$

In order to obtain a necessary condition for a relative minimum we shall apply Theorem 1.1. To this end, we embed an assumed minimum $x(t)$ in a one-parameter family

$$x_\varepsilon(t) = x(t) + \varepsilon\eta(t), \tag{1.7}$$

where $\eta(t) = (\eta^1(t), \ldots, \eta^n(t)) \in C_n^2$ and $\eta(a) = \eta(b) = \mathbf{0}$. Clearly, the $x_\varepsilon(t)$ are admissible functions in $A_n^2$. Furthermore, we note that (1.7) is short for the $n$ equations

$$x_\varepsilon^k(t) = x^k(t) + \varepsilon\eta^k(t) \qquad (k = 1, \ldots, n). \tag{1.8}$$

Now the task is to compute the first variation of $J$ at $x(t)$ and then apply Theorem 1.1.

**1.1   Lemma**   If $x(t)$ is a relative minimum of the functional $J$ defined by (1.6), then

$$\int_a^b \left( \frac{\partial L}{\partial x^k} - \frac{d}{dt}\frac{\partial L}{\partial \dot{x}^k} \right)\eta^k(t)\, dt = 0 \tag{1.9}$$

for all $\eta(t) \in C_n^2$ with $\eta(a) = \eta(b) = \mathbf{0}$.

Before presenting a proof of this lemma, we make some remarks concerning notation. First, the derivatives of $L$ which appear in the integrand of (1.9) are functions of $t$ and mean the following:

$$\frac{\partial L}{\partial x^k} \equiv \frac{\partial L}{\partial x^k}(t, x(t), \dot{x}(t)), \qquad \frac{\partial L}{\partial \dot{x}^k} \equiv \frac{\partial L}{\partial y^k}(t, x(t), \dot{x}(t)). \tag{1.10}$$

This is, the derivative of $L$ is taken with respect to the given argument and afterwards evaluated at the point $(t, x(t), \dot{x}(t))$. Second, on the left-hand side of (1.9) we are assuming the so-called "summation convention" of summing over a repeated index (in this case $k$, where $k$ ranges from 1 to $n$). We shall adhere strictly to this convention in the sequel unless explicitly stated otherwise.

*Proof of Lemma 1.1*    According to Theorem 1.1 we need to calculate $\delta J(x(t), \eta(t))$ and set it equal to zero. By Definition 1.1,

$$\delta J(x(t), \eta(t)) = \left( \frac{d}{d\varepsilon} \int_a^b L(t, x(t) + \varepsilon\eta(t), \dot{x}(t) + \varepsilon\dot{\eta}(t))\, dt \right)_{\varepsilon = 0}. \qquad (1.11)$$

By the regularity assumption on $L$, the derivative $d/d\varepsilon$ may be brought under the integral sign. An application of the chain rule and evaluation of the result at $\varepsilon = 0$ yields

$$\delta J = \int_a^b \left( \frac{\partial L}{\partial x^k} \eta^k(t) + \frac{\partial L}{\partial \dot{x}^k} \dot{\eta}^k(t) \right) dt.$$

The second term may be integrated by parts to obtain

$$\delta J = \int_a^b \left( \frac{\partial L}{\partial x^k} - \frac{d}{dt} \frac{\partial L}{\partial \dot{x}^k} \right) \eta^k(t)\, dt + \left. \frac{\partial L}{\partial \dot{x}^k} \eta^k(t) \right|_a^b.$$

Since $\eta^k(a) = \eta^k(b) = 0$ for $k = 1, \ldots, n$, we have

$$\delta J = \int_a^b \left( \frac{\partial L}{\partial x^k} - \frac{d}{dt} \frac{\partial L}{\partial \dot{x}^k} \right) \eta^k(t)\, dt,$$

which is the first variation of $J$. Since this variation must vanish at the minimum, we obtain the required result.    ☐

Lemma 1.1 gives a necessary condition for $x(t)$ to be a local minimum of the functional $J$; however, the condition (1.9) is not one which can be readily solved to determine $x(t)$, the latter being an obviously desirable approach to follow. Therefore, we seek to put (1.9) in a form more amenable to analysis. The so-called Fundamental Lemma of the calculus of variations provides the mechanism for this transition; it follows.

**1.2  Lemma**    Let $f(t)$ be a continuous real-valued function defined on $a \leq t \leq b$ and suppose that

$$\int_a^b f(t)\phi(t)\, dt = 0 \qquad (1.12)$$

for every $\phi \in C^2[a, b]$ satisfying $\phi(a) = \phi(b) = 0$. Then

$$f(t) \equiv 0, \qquad t \in [a, b].$$

*Proof*   By way of contradiction, assume there is a point $t_0 \in (a, b)$ for which $f(t_0) > 0$. By the continuity of $f$, there is an interval $(t_1, t_2)$ contained in $(a, b)$ about $t_0$ on which $f$ is strictly positive. Now let

$$\phi(t) = \begin{cases} (t - t_1)^3(t_2 - t)^3 & \text{for} \quad t \in (t_1, t_2), \\ 0 & \text{for} \quad t \notin (t_1, t_2). \end{cases}$$

Then $\phi \in C^2[a, b]$ and

$$\int_a^b f(t)\phi(t)\, dt = \int_{t_1}^{t_2} f(t)\phi(t)\, dt > 0,$$

a contradiction to (1.12). This completes the proof.   $\square$

Now we can obtain the following basic theorem.

**1.2   Theorem**   A necessary condition for the function $x(t) \in A_n^{\ 2}$ to provide a relative minimum for the functional $J$ defined by (1.6) is that its components $x^1(t), \ldots, x^n(t)$ satisfy the $n$ equations

$$\frac{\partial L}{\partial x^k} - \frac{d}{dt}\frac{\partial L}{\partial \dot{x}^k} = 0 \qquad (k = 1, \ldots, n) \tag{1.13}$$

for $a \le t \le b$.

*Proof*   Since the conclusion of Lemma 1.1 holds for all $\eta(t) \in C_n^{\ 2}$ with $\eta(a) = \eta(b) = \mathbf{0}$, in particular the conclusion holds when all of its components vanish except one, i.e., $\eta(t) = (0, \ldots, 0, \eta^i(t), 0, \ldots, 0)$, for some fixed $i$. Then (3.7) implies

$$\int_a^b \left( \frac{\partial L}{\partial x^i} - \frac{d}{dt}\frac{\partial L}{\partial \dot{x}^i} \right)\eta^i(t)\, dt = 0 \qquad (\text{no sum on } i)$$

for all $\eta^i(t)$ ($i$ fixed) of class $C^2(a, b)$ that vanish at $a$ and $b$. The regularity conditions on $L$ imply that the expression $\partial L/\partial x^i - d/dt(\partial L/\partial \dot{x}^i)$ is continuous on $[a, b]$, and therefore, by Lemma 1.2 it follows that

$$\frac{\partial L}{\partial x^i} - \frac{d}{dt}\frac{\partial L}{\partial \dot{x}^i} = 0.$$

But $i$ is arbitrary, so (1.13) follows.   $\square$

**1.1   Remark**   Equations (1.13) form a system of $n$ second-order ordinary differential equations which, in general, are nonlinear. They are called the *Euler–Lagrange equations* associated with the fundamental integral (1.6) (other authors call them just the *Euler equations*, and in physics they are known as *Lagrange's equations*). Theorem 1.2, which we emphasize gives

only a necessary condition for a relative minimum, tells us that the components $x^k(t)$ of a local minimum $x(t)$ must satisfy this system of differential
equations. However, we cannot expect that solutions to the Euler–Lagrange
equations will actually provide a local minimum. The situation is analogous
to that in the calculus where, after determining the critical points of a function
$y = f(x)$ via the necessary condition $f'(x) = 0$, we must then determine
which, if any, of the critical points lead to local minima. Whereas in the
calculus the second derivative test provides such conditions, in the calculus
of variations the so-called second variation plays a similar role. These
sufficient conditions, however, will be of no concern to us in the investigation
of the invariance properties of the given functional. So, in view of these
comments, any solution of the Euler–Lagrange equations will be called an
*extremal* and that solution is said to render the fundamental integral (1.6)
*stationary*.

**1.2  Remark**    In certain special cases, it is simple to find partial solutions
to the Euler–Lagrange differential equations in the form of a *first integral*,
or an expression

$$f(t, x^1, \ldots, x^n, \dot{x}^1, \ldots, \dot{x}^n)$$

that is constant on the extremal, i.e.,

$$f(t, x^1(t), \ldots, x^n(t), \dot{x}^1(t), \ldots, \dot{x}^n(t)) = \text{constant}$$

when $x^k = x^k(t)$, $k = 1, \ldots, n$, where $x(t)$ is an extremal. With no loss of
generality in the following, let us take $n = 1$.

(a)    If $L = L(t, \dot{x})$, i.e., if the Lagrangian does not depend explicitly
upon the coordinate $x$, then $\partial L/\partial x = 0$, and it follows directly from (1.13)
that

$$f(t, \dot{x}) \equiv \frac{\partial L}{\partial \dot{x}} = \text{constant}.$$

Solving for $\dot{x}$, we obtain an equation of the form $\dot{x} = F(t, c)$, from which
$x = x(t)$ can be found by a quadrature.

(b)    If $L = L(x, \dot{x})$, i.e., if the Lagrangian does not depend explicitly
upon $t$, then $\partial L/\partial t = 0$ and, as a result,

$$f(x, \dot{x}) \equiv L - \dot{x} \frac{\partial L}{\partial \dot{x}} = \text{constant}$$

along extremals. To observe this, we differentiate the expression on the left to obtain

$$\frac{d}{dt}\left(L - \dot{x}\frac{\partial L}{\partial \dot{x}}\right) = \frac{\partial L}{\partial x}\dot{x} + \frac{\partial L}{\partial \dot{x}}\ddot{x} - \left(\ddot{x}\frac{\partial L}{\partial \dot{x}} + \dot{x}\frac{d}{dt}\frac{\partial L}{\partial \dot{x}}\right)$$

$$= \dot{x}\left(\frac{\partial L}{\partial x} - \frac{d}{dt}\frac{\partial L}{\partial \dot{x}}\right)$$

$$= 0.$$

(c)  If $L = L(t, x)$, i.e., if the Lagrangian does not depend on the derivative $\dot{x}$, then $\partial L/\partial \dot{x} = 0$ and it follows from (1.13) that

$$\frac{\partial L}{\partial x}(t, x) = 0,$$

which is an algebraic equation rather than a differential equation.

**1.1  Example**  Among all curves of class $C^2$ joining the points $(t_1, x_1)$ and $(t_2, x_2)$ in the plane, determine the one that generates the surface of minimum area when rotated about the $t$-axis. From elementary calculus we know that the area of the surface of revolution generated by rotating the curve $x = x(t)$ about the $t$-axis is given by

$$J(x) = 2\pi \int_{t_1}^{t_2} x(t)\sqrt{1 + \dot{x}(t)^2}\ dt.$$

The Lagrangian $L(t, x, \dot{x}) = 2\pi x\sqrt{1 + \dot{x}^2}$ does not depend explicitly on $t$, so by Remark 1.2b it follows that the Euler–Lagrange equation yields

$$L - \dot{x}\frac{\partial L}{\partial \dot{x}} = \text{constant}$$

or in this case,

$$x\sqrt{1 + \dot{x}^2} - \dot{x}^2 x(1 + \dot{x}^2)^{-1/2} = C_1.$$

Solving for $\dot{x}$ we get

$$\frac{dx}{dt} \equiv \dot{x} = \frac{\sqrt{x^2 - C_1^{\ 2}}}{C_1^{\ 2}}.$$

Upon separating variables and integrating we obtain

$$t + C_2 = C_1 \log\left[\frac{x + \sqrt{x^2 - C_1^{\ 2}}}{C_1}\right],$$

which is the same as

$$x = C_1 \cosh\left(\frac{t + C_2}{C_1}\right).$$

Thus, the extremals are catenaries; the arbitrary constants of integration $C_1$ and $C_2$ can be determined by the boundary conditions $x(t_1) = x_1$ and $x(t_2) = x_2$.

**1.3   Remark**   The preceding discussion has of course left much unsaid. For example, the differentiability conditions imposed on the Lagrange function $L$ and the conditions on the admissible functions $x(t)$ can be relaxed considerably or constraints can be imposed on the problem. However, since these generalizations will play no role in our study of invariance criteria, we shall be content to work with the problem as formulated in this section, leaving these generalizations for the reader either to develop for himself or seek out in the literature.

**1.4   Remark**   There is a suggestive geometrical language which is helpful in understanding the basic variational problem as we have stated it. If we regard the set of $n + 1$ coordinates $t, x^1, \ldots, x^n$ as coordinates in an $(n + 1)$-dimensional space, the so-called *configuration space*, then the equations

$$x^k = x^k(t) \qquad (k = 1, \ldots, n, \qquad t \in [a, b]) \tag{1.14}$$

can be regarded as the parametric equations of a curve $C$ lying in this space which joins the points $(a, x^1(a), \ldots, x^n(a))$ and $(b, x^1(b), \ldots, x^n(b))$. Therefore, for the curve $C$ to afford an extreme value to the fundamental integral, it is necessary that the functions (1.14) defining $C$ are such that the $n$ expressions

$$E_k \equiv \frac{\partial L}{\partial x^k} - \frac{d}{dt}\frac{\partial L}{\partial \dot{x}^k} \tag{1.15}$$

vanish along $C$. These expressions, which are the left-hand sides of the Euler equations, are called the *Euler expressions*.

## 1.4   APPLICATIONS TO CLASSICAL DYNAMICS

According to the doctrine of classical dynamics, one associates with the system being described a set of quantities or dynamical variables, each of which is a well-defined value at every instant of time, and which define the dynamical state of the system at that instant. Moreover, it is assumed that classical dynamics is deterministic; i.e., the time evolution of the system is completely determined if its state is given at an initial instant. Analytically, this doctrine is expressed by the fact that the dynamical variables satisfy a set of differential equations as functions of time (Newton's second law) along with initial conditions. The program of classical dynamics, therefore, consists of listing the dynamical variables and discovering the governing equations of motion which predict the system's evolution in time.

To be a little more precise, let $x^1, \ldots, x^n$ denote the generalized coordinates of a dynamical system. That is, regarded as functions of time, we assume that $x^1, \ldots, x^n$ completely specify the state of the system at any given time, and there exist no relations among the $x^k$ so that they may be regarded as independent. We further assume that there exists a state function $L = L(t, x^1, \ldots, x^n, \dot{x}^1, \ldots, \dot{x}^n)$ of class $C^2$ such that the governing equations of motion for the system are

$$\frac{\partial L}{\partial x^k} - \frac{d}{dt} \frac{\partial L}{\partial \dot{x}^k} = 0 \qquad (k = 1, \ldots, n).$$

The state function, or Lagrangian, $L$ can be regarded as a function which contains all the information about the system. Once it is known, then the equations which govern the evolution of the system can be written down directly and from them and the initial conditions, the trajectory of the system can be found. For many dynamical systems, the function $L$ is given by the difference between the kinetic and potential energies:

$$L(t, x, \dot{x}) = \tfrac{1}{2} g_{ij}(x) \dot{x}^i \dot{x}^j - V(x), \tag{1.16}$$

where the kinetic energy is a positive definite quadratic form in the generalized velocities $\dot{x}^j$, and $V(x)$ denotes the potential of the external forces acting on the system.

There is an alternate axiomatic approach to classical dynamics which fits well with the concepts of the calculus of variations. Such an approach is based upon the idea that a system should evolve along the path of "least resistance." Principles of this sort have a long history in physical theories dating back to the Greeks when Hero of Alexander stated a minimum principle concerning the paths of reflected light rays. In the seventeenth century, Fermat announced that light travels through a medium along the path of shortest time. For mechanical systems, Maupertuis' principle of least action stated that a system should evolve from one state to another such that the "action" (a vaguely defined term with the units energy × time) is smallest. Lagrange and Gauss were also adherents of similar minimum principles. However, in the early part of the nineteenth century William Rowan Hamilton stated what has become an encompassing, aesthetic principle which can be generalized to embrace many areas of physics.

Hamilton's principle roughly states that the evolution in time of a mechanical system takes place in such a manner that integral of the difference between the kinetic and potential energies for the system is stationary. Again, let $x^1, \ldots, x^n$ be generalized coordinates as described above, and suppose that the Lagrangian $L$ is given by (1.16). Hamilton's principle may be formulated for these systems by stating that the motion of the system from time $t_1$ to time $t_2$ is such that

$$J(x) = \int_{t_1}^{t_2} L(t, x(t), \dot{x}(t)) \, dt$$

is stationary for the curve $x = x(t) = (x^1(t), \ldots, x^n(t))$ in configuration space along which the motion takes place. In still different words, of all possible paths in configuration space joining the two points $(t_1, x(t_1))$ and $(t_2, x(t_2))$, the motion takes place along the curve $C$ which affords an extreme value to the fundamental integral or action integral. Consequently, from Theorem 1.2 it follows that if $x = x(t)$ is the path of the motion, then the components $x^k = x^k(t)$ of $C$ must satisfy the Euler–Lagrange equations

$$\frac{\partial L}{\partial x^k} - \frac{d}{dt} \frac{\partial L}{\partial \dot{x}^k} = 0 \qquad (k = 1, \ldots, n).$$

These are the governing equations, or equations of motion for the system. It is standard procedure to state Hamilton's principle in the form

$$\delta \int_{t_1}^{t_2} L(t, x(t), \dot{x}(t)) \, dt = 0.$$

However, care must be exercised in interpreting this latter statement as it has been formulated above.

**1.2  Example**   Consider the motion of a mass $m$ attached to a spring with stiffness $k$ (see Fig. 1). Its position measured from equilibrium is denoted

**FIGURE 1**

by the single generalized coordinate $x$. To be found, given an initial displacement $x_0$ and initial velocity $v_0$, is $x(t)$, the position as a function of time. From elementary physics, its kinetic energy is

$$T = \tfrac{1}{2} m \dot{x}^2$$

and its potential energy is

$$V = \tfrac{1}{2} k x^2 \qquad (k > 0).$$

Therefore, the Lagrangian for this system is

$$L = T - V = \tfrac{1}{2}m\dot{x}^2 - \tfrac{1}{2}kx^2,$$

and Hamilton's principle states that $x = x(t)$ must be such that

$$\delta \int_{t_1}^{t_2} (\tfrac{1}{2}m\dot{x}^2 - \tfrac{1}{2}kx^2)\, dt = 0.$$

Hence $x(t)$ must satisfy the Euler–Lagrange equation

$$\frac{\partial L}{\partial x} - \frac{d}{dt}\frac{\partial L}{\partial \dot{x}} = 0$$

or, in this case,

$$-kx - \frac{d}{dt}(m\dot{x}) = 0.$$

We notice that this equation, the equation which governs the motion, just expresses the fact that the restoring force $-kx$ on the mass $m$ equals the time rate of change of the linear momentum $m\dot{x}$; this is Newton's second law. The general solution to the equation

$$\ddot{x} + \frac{k}{m}x = 0,$$

which is the same as the one above, is given by

$$x(t) = C_1 \cos\sqrt{\frac{k}{m}}\, t + C_2 \sin\sqrt{\frac{k}{m}}\, t.$$

The arbitrary constants $C_1$ and $C_2$ can be determined from the initial conditions

$$x(t_0) = x_0, \qquad \dot{x}(t_0) = v_0.$$

We observe that the mass oscillates back and forth in simple harmonic motion, which follows from the form of the general solution given above.

**1.3  Example**  Consider the problem of determining the motion of $n$ bodies. In this case the Lagrangian takes the form $L = T - V$, where

$$T = \frac{1}{2}\sum_{i=1}^{n} m_i(\dot{x}_i{}^2 + \dot{y}_i{}^2 + \dot{z}_i{}^2) \tag{1.17}$$

and

$$V = \sum_{1 \le i < j \le n} G\left(\frac{m_i m_j}{r_{ij}}\right) \tag{1.18}$$

are the kinetic and potential energies of the system, and $m_i$ is the mass of $i$th body with position $(x_i, y_i, z_i)$; $G$ is the gravitational constant and

$$r_{ij} = \sqrt{(x_i - x_j)^2 + (y_i - y_j)^2 + (z_i - z_j)^2} \qquad (1.19)$$

is the distance between the $i$th and $j$th bodies. For this system there are $3n$ generalized coordinates $x_1, y_1, z_1, \ldots, x_n, y_n, z_n$. Hamilton's principle states that the system must evolve in such a way that the action integral $\int_{t_2}^{t_1} L\, dt$ is stationary. Lagrange's equations for this system again coincide with Newton's second law and are given by

$$\frac{\partial V}{\partial x_i} = \frac{d}{dt}(m_i \dot{x}_i),$$

$$\frac{\partial V}{\partial y_i} = \frac{d}{dt}(m_i \dot{y}_i), \qquad (1.20)$$

$$\frac{\partial V}{\partial z_i} = \frac{d}{dt}(m_i \dot{z}_i),$$

where $i = 1, \ldots, n$.

One might well infer from the previous examples that the practical value of Hamilton's principle is questionable. After all, it results in writing down the equations of motion (Newton's second law or Lagrange's equations); why not just directly determine the governing equations and forego the variation principle altogether? Actually, this is a legitimate objection, particularly in view of the fact that in much of the literature the variation principle is an a posteriori notion; i.e., oftentimes the variation principle for a given system is determined from the known equations of motion, and not conversely as would be relevant from viewpoint of the calculus of variations. Moreover, if a variation principle is enunciated as the basic tenet for the system, then there are complicated sufficient conditions that must be dealt with; the latter, however, seem to have little or no role in physical problems. And finally, although variation principles do to some extent represent a unifying concept for physical theories, the extent is by no means universal; for instance, it is oftentimes impossible to state such a principle for dissipative systems or systems having nonholonomic constraints. Therefore, in summary, there are definite objections that can be raised with respect to the *ab initio* formulation of a physical theory by a variation principle.

Even in the study of invariance criteria, which is the topic of our interest, it is the Lagrange function and its form which forces invariance upon us; as we shall observe, the action is invariant because of the invariance properties of the Lagrangian. This points out again the important role of the state function or Lagrangian for the system. It involves the internal properties of the system like the quantities of position, velocity, and energy (in contrast to the Newtonian formulation which involves external properties like force); it contains the information for the system—from it follows the equations of motion, and from its invariance properties follow important identities and conservation laws.

In spite of the objections raised above, we shall follow custom and not ignore the action integral. There are arguments on its side, too. The action integral plays a fundamental role in the development of numerical methods for solving differential equations (Rayleigh–Ritz method, Galerkin methods, etc.); it also plays a decisive role in the definition of Hamilton's characteristic function, the basis of the Hamilton–Jacobi canonical theory. Finally, many variation problems arise from problems in geometry or other areas apart from classical mechanics and physics; in these problems, the fundamental integral *is* an a priori notion. Therefore, in our study of invariance problems, we shall maintain the role of the functional or fundamental integral.

The purpose of this section was not to present a deep or thorough exposition on classical dynamics, but rather to comment on the role of the calculus of variations in physical problems. In the sequel, this role will become even more apparent since many of our applications to the general theory will lie in the realm of physics.

## 1.5   MULTIPLE INTEGRAL PROBLEMS

Whereas variation problems given by single integrals enjoy significant application in the formulation of fundamental principles of mechanical systems and geometry, variation problems defined by multiple integrals are important in the formulation of principles in the theory of physical fields. In this section we shall formulate these variation problems and write down the Euler–Lagrange equations for the multiple integral case. Again, our approach is based upon the general method outlined in Section 2 of this chapter.

In order to fix the notion of the region of integration, we shall let $D$ denote a closed rectangle or cylinder in $R^m$.† A point of $D$ will be denoted by $t =$

---

† More generally, we could take $D$ to be a compact, simply connected region in $R^m$ whose boundary, denoted by Bd $D$, is composed of finitely many $(m - 1)$-dimensional hypersurfaces which intersect in such a manner that only finitely many corners and edges appear and which precludes point boundaries and infinitely sharp edges. Such a region is called *normal*. A rigorous definition of normal regions is given by Klötzler [1]; such regions have the property that they are quite general and the Divergence Theorem holds over them, the latter property being needed for a derivation of the Euler–Lagrange equations in several dimensions. In our analysis, however, this degree of generality will not be required.

$(t^1, \ldots, t^m)$ and we let $dt = dt^1 \cdots dt^m$. The boundary of $D$ will be denoted by Bd $D$. The Divergence Theorem, which states that the volume integral of a divergence expression equals a certain integral over the Bd $D$, is needed in our calculation of necessary conditions, so we shall now state it but without proof. Any standard advanced calculus text can be consulted for the proof.

**1.3   Theorem**   Let $D \subseteq R^m$ and let $f(t) = f(t^1, \ldots, t^m)$ be continuous in $D$ with bounded, continuous first partial derivatives in $D$. Then

$$\int_D \frac{\partial f}{\partial t^\alpha} \, dt = \int_{Bd\ D} f(t) \cos(\mathbf{n}, t^\alpha) \, d\sigma,$$

where $\mathbf{n}$ denotes the outer oriented unit normal to Bd $D$, $d\sigma$ is a surface element of Bd $D$, and $\cos(\mathbf{n}, t^\alpha)$ denotes the cosine of the angle between $\mathbf{n}$ and the $t^\alpha$ coordinate.   $\square$

We now define the function spaces on which the variation problems will be defined. Let $C^2(D)$ denote the set of all continuous functions on $D$ whose second partial derivatives are continuous, and let $C_n{}^2(D)$ denote the set of all vector functions

$$x(t) = (x^1(t), \ldots, x^n(t))$$

on $D$ whose components $x^k(t)$ belong to $C^2(D)$. The class of admissible functions are those functions in $C_n{}^2(D)$ which take prescribed values on Bd $D$; i.e.,

$$A_n{}^2(D) = \{x(t) \in C_n{}^2(D) : x(t) = \phi(t),\ t \in Bd\ D,\ \phi\ \text{given on Bd}\ D\}. \quad (1.21)$$

Geometrically, we have the following interpretation. If $t^1, \ldots, t^m; x^1, \ldots, x^n$ denote the coordinates in $R^{m+n}$, and if $x(t) \in A_n{}^2(D)$, then the equations

$$x^k = x^k(t), \qquad t \in D \qquad (k = 1, \ldots, n)$$

represent an $m$-dimensional hypersurface $C_m$ in $R^{m+n}$ whose boundary is defined by

$$x^k = \phi^k(t), \qquad t \in Bd\ D \qquad (k = 1, \ldots, n),$$

where $\phi(t) = (\phi^1(t), \ldots, \phi^n(t))$ is a given function defined on the boundary of $D$. When $m = 2$ and $n = 1$ there is a simple geometric interpretation as shown in Fig. 2. In this case the boundary of $C_2$ is a closed curve $\Gamma$ which is the image of Bd $D$ under the fixed mapping $\phi = \phi(t^1, t^2)$.

We now define the variation problem that we shall consider. Let $L = L(t^1, \ldots, t^m; x^1, \ldots, x^n; y^1, \ldots, y^{mn})$ be a real-valued function defined

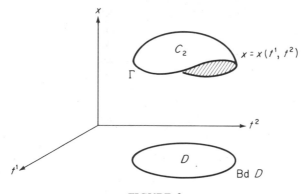

**FIGURE 2**

on $R^{m+n+mn}$ which is twice continuously differentiable in each of its arguments, and let $x(t) = (x^1(t), \ldots, x^n(t)) \in A_n^2(D)$. We denote the first and second partial derivatives of the components of $x(t)$ by

$$\dot{x}_\alpha^k = \frac{\partial x^k}{\partial t^\alpha}, \qquad \ddot{x}_{\alpha\beta}^k = \frac{\partial^2 x^k}{\partial t^\beta \, \partial t^\alpha}. \tag{1.22}$$

The $m$-fold *fundamental integral* is then defined on $A_n^2(D)$ by

$$J(x(t)) = \int_D L(t^1, \ldots, t^m; x^1(t), \ldots, x^n(t);$$

$$\dot{x}_1^{\,1}(t), \ldots, \dot{x}_1^{\,n}(t); \ldots; \dot{x}_m^{\,1}(t), \ldots, \dot{x}_m^{\,n}(t)) \, dt, \tag{1.23}$$

where $dt = dt^1 \cdots dt^m$. More concisely, we write (1.23) as

$$J(x(t)) = \int_D L\left(t, x(t), \frac{\partial x(t)}{\partial t}\right) dt, \tag{1.24}$$

where $\partial x(t)/\partial t$ denotes the set of all first partials $\partial x^k(t)/\partial t^\alpha$, $k = 1, \ldots, n$; $\alpha = 1, \ldots, m$. We note that the admissible functions form a subset of $C_n^2(D)$ which is a normed linear space with the norm given by

$$\|x(t)\| = \max_D \{|x^1(t)|, \ldots, |x^n(t)|\} + \max_D \left\{ \left| \frac{\partial x^k}{\partial t^\alpha} \right| \right\}. \tag{1.25}$$

Therefore, we have the variation problem

$$\int_D L\left(t, x(t), \frac{\partial x(t)}{\partial t}\right) dt \to \min \qquad \text{on} \quad A_n^2(D).$$

Following the general approach outlined in Section 2, let $x(t)$ be a local minimum; i.e., suppose that for some $\delta > 0$ we have $J(x(t)) \le J(y(t))$ for all

$y(t) \in A_n^2(D)$ with $\|x(t) - y(t)\| < \delta$. Consider the one-parameter family $x(t, \varepsilon)$, $|\varepsilon| < \varepsilon_0(\delta)$, of functions of the form

$$x(t, \varepsilon) = x(t) + \varepsilon \eta(t), \qquad (1.26)$$

where $\eta(t) \in C_n^2(D)$, $\eta(t) = \mathbf{0}$ whenever $t \in \operatorname{Bd} D$, and $\|x(t, \varepsilon) - x(t)\| < \delta$. Then for $|\varepsilon| < \varepsilon_0$ we have $x(t, \varepsilon) \in A_n^2(D)$. Equation (1.26) is of course short for the $n$ equations

$$x^k(t, \varepsilon) = x^k(t) + \varepsilon \eta^k(t),$$

where $\eta(t) = (\eta^1(t), \ldots, \eta^n(t))$ and $x(t, \varepsilon) = (x^1(t, \varepsilon), \ldots, x^n(t, \varepsilon))$.

Next we compute the first variation $\delta J(x(t), \eta(t))$ and apply Theorem 1.1. The result is summarized in the following lemma, which is the higher dimensional analog of Lemma 1.1.

**1.3 Lemma** If $x(t) \in A_n^2(D)$ is a local minimum of the functional (1.24), then

$$\int_D \left( \frac{\partial L}{\partial x^k} - \frac{\partial}{\partial t^\alpha} \frac{\partial L}{\partial \dot{x}_\alpha^{\ k}} \right) \eta^k(t) \, dt = 0 \qquad (1.27)$$

for all $\eta(t) \in C_n^2(D)$ which vanish on $\operatorname{Bd} D$.

Again we remark that in (1.27) the derivatives of $L$ which occur in the integrand are shorthand notation for more complicated expressions, for example,

$$\frac{\partial L}{\partial x^k} \equiv \frac{\partial L}{\partial x^k}\left( t, x(t), \frac{\partial x(t)}{\partial t} \right).$$

Also, the summation convention is assumed. Here, and in the sequel, lower case Latin letters $i, j, k, \ldots$ will range over $1, \ldots, n$ and lower case Greek letters $\alpha, \beta, \ldots$ will range over $1, \ldots, m$. To get on with the proof of Lemma 1.3, we appeal to Definition 1.1 and Theorem 1.1 to obtain

$$\delta J(x(t), \eta(t)) = \left( \frac{d}{d\varepsilon} \int_D L\left( t, x(t) + \varepsilon \eta(t), \frac{\partial x(t)}{\partial t} + \varepsilon \frac{\partial \eta(t)}{\partial t} \right) dt \right)_{\varepsilon = 0} = 0 \quad (1.28)$$

for all $\eta(t)$ satisfying the conditions of the lemma. Here, of course, $\partial x(t)/\partial t + \varepsilon(\partial \eta(t)/\partial t)$ is a concise form for the $mn$ expressions

$$\frac{\partial x^k}{\partial t^\alpha} + \varepsilon \frac{\partial \eta^k}{\partial t^\alpha}.$$

Since both $L$ and $\partial L/\partial \varepsilon$ are continuous functions of $\varepsilon$ and $t$, the derivative may be brought inside the integral in (1.28), and then an application of the chain rule yields

$$\int_D \left( \frac{\partial L}{\partial x^k} \eta^k(t) + \frac{\partial L}{\partial \dot{x}_\alpha^{\ k}} \dot{\eta}_\alpha^{\ k} \right) dt = 0. \tag{1.29}$$

Using the identity

$$\frac{\partial}{\partial t^\alpha} \left( \frac{\partial L}{\partial \dot{x}_\alpha^{\ k}} \eta^k \right) = \frac{\partial L}{\partial \dot{x}_\alpha^{\ k}} \dot{\eta}_\alpha^{\ k} + \frac{\partial}{\partial t^\alpha} \left( \frac{\partial L}{\partial \dot{x}_\alpha^{\ k}} \right) \eta^k, \tag{1.30}$$

we observe that (1.29) may be rewritten as

$$\int_D \left( \frac{\partial L}{\partial x^k} - \frac{\partial}{\partial t^\alpha} \frac{\partial L}{\partial \dot{x}_\alpha^{\ k}} \right) \eta^k(t)\, dt + \int_D \frac{\partial}{\partial t^\alpha} \left( \frac{\partial L}{\partial \dot{x}_\alpha^{\ k}} \eta^k \right) dt = 0. \tag{1.31}$$

The functions $(\partial L/\partial \dot{x}_\alpha^{\ k})\eta^k$ are continuously differentiable on $D$ and therefore satisfy the hypotheses of the Divergence Theorem; consequently, upon applying Theorem 1.3, Eq. (1.31) becomes

$$\int_D \left( \frac{\partial L}{\partial x^k} - \frac{\partial}{\partial t^\alpha} \frac{\partial L}{\partial \dot{x}_\alpha^{\ k}} \right) \eta^k(t)\, dt + \int_{\mathrm{Bd}\, D} \frac{\partial L}{\partial \dot{x}_\alpha^{\ k}} \eta^k(t) \cos(\eta, t^\alpha)\, d\tau = 0.$$

The integral over the boundary vanishes since $\eta^k(t) = 0$ on Bd $D$, and thus the lemma is proved.   □

In order to cast (1.27) in a more useful form, we require the following lemma which is another form of the Fundamental Lemma (Lemma 1.2).

**1.4   Lemma**   If $f(t)$ is continuous and real-valued on $D$ and if

$$\int_D f(t)h(t)\, dt = 0 \tag{1.32}$$

for every $h(t) \in C^2(D)$ which vanishes on Bd $D$, then

$$f(t) \equiv 0 \qquad \text{for all} \quad t \in D.$$

*Proof*   On the contrary, assume there is $t_0$ in the interior of $D$ for which $f(t_0) > 0$. Then, by the continuity of $f$, there is a sphere $S: |t - t_0| < \delta$ in $D$ of radius $\delta > 0$ and centered at $t_0$ for which $f(t) > 0$ for $t \in S$. Now, define on $D$ the function

$$h(t) = \begin{cases} 0, & |t - t_0| \geq \delta, \\ (|t - t_0|^2 - \delta^2)^4, & |t - t_0| < \delta. \end{cases}$$

Then $h(t) \in C^2(D)$ and $h(t) = 0$ on Bd $D$; therefore

$$\int_D f(t)h(t)\, dt = \int_S f(t)h(t)\, dt > 0$$

since both $f$ and $h$ are positive on $S$. This fact provides the needed contradiction to the hypothesis (1.32).  $\square$

By combining Lemmas 1.3 and 1.4 as in Section 3 we are now in position to obtain an important necessary condition for a local minimum.

**1.4  Theorem**  If $x(t) \in A_n^2(D)$ is a local minimum of the functional $J$ defined by (1.24), then the components $x^k(t)$ of $x(t)$ must satisfy the $n$ equations

$$\frac{\partial L}{\partial x^k} - \frac{\partial}{\partial t^\alpha}\frac{\partial L}{\partial \dot{x}_\alpha^k} = 0 \qquad (k = 1, \ldots, n) \tag{1.33}$$

identically in $D$.

*Proof*  Let us denote

$$E_k(t) \equiv \frac{\partial L}{\partial x^k} - \frac{\partial}{\partial t^\alpha}\frac{\partial L}{\partial \dot{x}_\alpha^k} \qquad (k = 1, \ldots, n).$$

By Lemma 1.3

$$\int_D E_k(t)\eta^k(t)\, dt = 0$$

for all $\eta(t) = (\eta^1(t), \ldots, \eta^n(t)) \in C_n^2(D)$ with $\eta(t) \equiv 0$ on Bd $D$. Set $\eta^k(t) = 0$ for $k \ne i$, where $i$ is a fixed index between 1 and $n$. Then

$$\int_D E_i(t)\eta^i(t)\, dt = 0 \qquad \text{(no sum on } i)$$

for all possible $\eta^i(t) \in C^2(D)$ with $\eta^i(t) = 0$ on Bd $D$. Since $E_i(t)$ is continuous on $D$ it follows from Lemma 1.4 that $E_i(t) = 0$ on $D$. Letting $i = 1, \ldots, n$, we obtain the set of equations (1.33).  $\square$

**1.5  Remark**  Equations (1.33) are known as the Euler–Lagrange equations for the variation problem defined by the multiple integral (1.24). These equations are, in general, a set of $n$ second-order nonlinear partial differential equations. Written out in full, they become

$$\frac{\partial^2 L}{\partial t^\alpha\, \partial \dot{x}_\alpha^k} - \frac{\partial L}{\partial x^k} + \frac{\partial^2 L}{\partial x^j\, \partial \dot{x}_\alpha^k}\dot{x}_\alpha^j + \frac{\partial^2 L}{\partial \dot{x}_\beta^i\, \partial \dot{x}_\alpha^k}\ddot{x}_{\alpha\beta}^i = 0. \tag{1.34}$$

Any solution of (1.33), or equivalently (1.34), is called an *extremal surface,* and the *n* expressions

$$E_k \equiv \frac{\partial L}{\partial x^k} - \frac{\partial}{\partial t^\alpha} \frac{\partial L}{\partial \dot{x}_\alpha{}^k}$$

will be called the *Euler expressions* corresponding to the multiple integral problem.

**1.4 Example** (*Plateau's Problem*) *To be found is the surface of least area whose boundary is a fixed contour.* Using Fig. 2 as our geometrical model, we seek the function $x = x(t^1, t^2)$ describing the surface $C_2$ whose boundary is the fixed closed curve $\Gamma$. From elementary calculus, the surface area is given by the functional

$$J(x) = \int_D \sqrt{1 + \dot{x}_1{}^2 + \dot{x}_2{}^2}\, dt^1\, dt^2,$$

where $D$ is the region in the $t^1 t^2$-plane over which the surface lies, and $\dot{x}_1 = \partial x/\partial t^1$, $\dot{x}_2 = \partial x/\partial t^2$, as dictated by our earlier notational convention. The Euler–Lagrange equation for this problem is given by

$$\frac{\partial L}{\partial x} - \frac{\partial}{\partial t^1} \frac{\partial L}{\partial \dot{x}_1} - \frac{\partial}{\partial t^2} \frac{\partial L}{\partial \dot{x}_2} = 0$$

which, when substitutions and simplifications are made, yields

$$\ddot{x}_{11}(1 + \dot{x}_2{}^2) - 2\dot{x}_1\dot{x}_2\ddot{x}_{12} + \ddot{x}_{22}(1 + \dot{x}_1{}^2) = 0.$$

It can be shown that this nonlinear second-order partial differential equation implies that the mean curvature of the minimal surface vanishes. Explicit solutions are difficult or impossible to obtain.

Multiple integral problems in the calculus of variations also have vast application to field theory in physics. In this case we may think of the functions $x^k(t)$ as "field" functions which describe the behavior of the physical system, where $t = (t^1, \ldots, t^4)$ represent space-time coordinates. If the Lagrangian $L$ is properly chosen, then the Euler–Lagrange equations can be made to coincide with the so-called "field equations" for the system. Thus, we obtain a variation principle for the field. For example, when the $x^k(t)$ are given by the 4-potential of the electromagnetic field (in a vacuum, say), a Lagrangian can be found whose associated Euler–Lagrange equations coincide with Maxwell's equations, the governing equations of electromagnetism. Except for this comment, we shall postpone a discussion of field theory to a later chapter.

## 1.6   INVARIANCE—A PREVIEW

As we stated in the introduction, our primary goal is to investigate the invariance properties of the fundamental integrals which occur in the calculus of variations. We are now in a position to give some brief remarks concerning the nature of these problems and indicate the direction which we shall follow.

The study of invariance problems was initiated in the early part of this century by Emmy Noether in a classical paper on the subject [1]. Influenced by the work of Klein [1] and of Lie [1] on the transformation properties of differential equations under continuous groups of transformations, Noether proved two fundamental results, now known as the Noether theorems; classically, they can be stated as follows (more precise versions will be given later):

(I)   If the fundamental integral $J$ is invariant under an $r$-parameter continuous group of transformations of the variables, then there result $r$ identities between the Euler–Lagrange expressions $E_k$ and quantities which can be written as divergences.

(II)   If the fundamental integral $J$ is invariant under a group of transformations which depend upon $\tau$ arbitrary functions and their derivatives up to some order $\sigma$, then there exist identities between the Euler–Lagrange expressions $E_k$ and their derivatives up to order $\sigma$.

Physically, the Noether identities have important consequences. For example, if $L$ is the Lagrangian of some physical system, then the invariance of the fundamental integral under a $r$-parameter group leads directly to conservation laws for the system. In the single integral case this means that the invariance hypothesis leads to expressions which are constant along the extremals, or in other words, first integrals of the equations of motion. For multiple integral problems, the conservation laws take the form of a vanishing divergence which, as we shall see, can be interpreted in the usual way as conservation of a "flux" quantity. The second theorem, which is not as well known as the first, is related to parameter invariant variation problems, i.e., variation integrals which are invariant under arbitrary transformations of the independent variables. This type of invariance requirement, when imposed on an arbitrary variation integral, restricts the form of the Lagrangian severely.

Our main concern in this work is variation problems which admit a finite continuous group, i.e., a transformation group depending upon finitely many parameters. Our approach to these problems is somewhat nonstandard in the following manner. The usual approach to the first Noether theorem is to derive the theorem as a corollary of the so-called *fundamental variational formula* which expresses the first variation of the fundamental integral when both the dependent and independent variables are subjected to *small variations*. The approach that we follow in this work is more direct and we completely avoid the method of small variations. This direct method is fruitful in several respects. Besides being simpler, the method leads to a set of fundamental invariance identities which follow as necessary conditions for the fundamental integral to be invariant under a finite continuous group. These invariance identities, in addition to providing simple access to Noether's theorem, have inherent value in their own right. For example, these invariance identities enable us to characterize classes of Lagrangians which possess given invariance properties under an invariance group. Moreover, these identities lead to conditions which allow us to determine a transformation group under which a given fundamental integral is invariant and thereby provide a method for determining first integrals of the governing equations of motion via Noether's theorem.

## 1.7   BIBLIOGRAPHIC NOTES

Our treatment in this chapter of the rudiments of the calculus of variations has been, by necessity, limited in motivation and scope. Several excellent treatments of the problems and theory of the calculus of variations can be consulted by those interested in pursuing an in-depth approach. Among some of the more modern textbooks are those by Akhiezer [1], Elsgolc [1], Epheser [1], Funk [1], Gelfand and Fomin [1], Sagan [1], and Weinstock [1]. Also, the classical treatment of Bolza [1, 2] still stands as a monument in the subject, and the chapter in Courant and Hilbert [1] is outstanding. For multiple integral problems we mention Klötzler [1] and Morrey [1], and the theory treated from the viewpoint of the Hamilton–Jacobi Theory can be found in Rund [1]. Bliss [1] has written an elementary, popular account of the subject which also can be used as an initial source for historical information.

Concerning applications, nearly any book on advanced dynamics will have some information relating the calculus of variations to problems in mechanics and on Hamilton's principle. In particular, we mention Goldstein [1] and Marion [1]. The treatise by Pars [1] and the classic work by Lanczos [1] have a lot of information on this subject. For variational principles in physical fields, a subject which we shall discuss in Chapter 5, we mention the books by Barut [1], Bogoluibov and Shirkov [1], Landau and Lifschitz [1], Rohrlich [1], and W. R. Davis [1]. Variational principles in the theory of relativity are discussed in Misner, Thorne, Wheeler [1] and Landau and Lifschitz [1].

## EXERCISES

**1-1**  Consider the variational problem $J = \int_0^{2\pi} \dot{x}^2 \, dt$. Also consider the function $x(t) = t$ and the "varied" family $x_\varepsilon(t) = t + \varepsilon \sin t$. Sketch this family geometrically, compute $J(\varepsilon)$ explicitly, and prove directly that $\delta J(x, \eta) = 0$, where $x = t, \eta = \sin t$. Deduce that $x(t) = t$ is an extremal.

**1-2**  For the simplest variational problem $J = \int_{t_0}^{t_1} L(t, x, \dot{x}) \, dt$, prove that the Euler–Lagrange equations may be written in the form

$$\frac{\partial L}{\partial x^k} - \frac{d}{dt}\left(L - \dot{x}^k \frac{\partial L}{\partial \dot{x}^k}\right) = 0.$$

**1-3**  Analyze the following variational problems:

(a)  $J(x) = \displaystyle\int_0^1 \dot{x} \, dt \to \text{Ext}, \qquad x(0) = 0, \quad x(1) = 1.$

(b)  $J(x) = \displaystyle\int_0^1 x\dot{x} \, dt \to \text{Ext}, \qquad x(0) = 0, \quad x(1) = 1.$

(c)  $J(x) = \displaystyle\int_0^1 tx\dot{x} \, dt \to \text{Ext}, \qquad x(0) = 0, \quad x(1) = 1.$

**1-4**  Find extremals of the functional

$$J(x) = \int_a^b (x^2 + \dot{x}^2 + 2xe^t) \, dt.$$

**1-5**  Calculate the first variation of the functional

$$J(x) = \int_0^1 \int_0^1 K(s, t)x(t)x(s) \, ds \, dt.$$

**1-6**  Find extremals of the functional

$$J(x) = \int_a^b f(t)\sqrt{1 + \dot{x}^2} \, dt,$$

where $f(t)$ is a given $C^2$ function.

**1-7**  Find the Euler–Lagrange equations corresponding to the following functionals:

(a)  $J(u) = \int_D \left\{ \left(\dfrac{\partial u}{\partial x}\right)^2 + \left(\dfrac{\partial u}{\partial y}\right)^2 + 2u\rho \right\} dx\,dy$, where $\rho = \rho(x, y)$.

(b)  $J(u) = \int_D \left\{ x^2\left(\dfrac{\partial u}{\partial x}\right)^2 + y^2\left(\dfrac{\partial u}{\partial y}\right)^2 \right\} dx\,dy$.

(c)  $J(u) = \int_D \dfrac{1}{2}\left( g^{\alpha\beta}\, \dfrac{\partial u}{\partial t^\alpha}\, \dfrac{\partial u}{\partial t^\beta} + m^2 u^2 \right) dt^1 \cdots dt^4$, where

$$g^{\alpha\beta} = \begin{bmatrix} -1 & 0 & 0 & 0 \\ 0 & 1 & 0 & 0 \\ 0 & 0 & 1 & 0 \\ 0 & 0 & 0 & 1 \end{bmatrix} \qquad \text{and } m \text{ is a constant.}$$

**1-8**  Prove that the functional

$$J(x) = \int_0^1 (\dot{x} \sin \pi x - (t + x)^2)\, dt$$

*assumes* its maximum value $2/\pi$ for the function $x(t) = -t$.

**1-9**  Find extremals of the variational problem

$$J(x^1, x^2) = \int_0^{\pi/2} ((\dot{x}^1)^2 + (\dot{x}^2)^2 + 2x^1 x^2)\, dt \to \text{Ext}$$

with

$$x^1(0) = 0, \qquad x^1\left(\frac{\pi}{2}\right) = 1, \qquad x^2(0) = 0, \qquad x^2\left(\frac{\pi}{2}\right) = 1.$$

**1-10**  Let

$$J(x) = \exp \int_0^1 x(t)^2\, dt$$

be a functional defined on $C[0, 1]$. Calculate its first variation and determine a necessary condition for a minimum.

**1-11**  For the simplest variation problem $(n = 1)$

$$J(x) = \int_a^b L(t, x, \dot{x})\, dt,$$

determine the form of the most general Lagrangian for which the Euler–Lagrange equation is satisfied identically.

**1-12** (Brachistochrone Problem)   A bead of mass $m$ with initial velocity zero falls with no friction under the force of gravity $g$ from a point $(x_1, y_1)$ to a point $(x_2, y_2)$ along a wire defined by a curve $y = y(x)$ in the $xy$-plane $(x_1 < x_2, y_1 > y_2)$. Show that the time required for the bead to make its passage is given by

$$J(y) = \int_{x_0}^{x_1} \frac{\sqrt{1 + y'^2}}{\sqrt{2g(y_1 - y)}} \, dx.$$

Prove that the path giving the shortest transit time (the brachistochrone) is an arc of a cycloid, which can be written parametrically as

$$x = \frac{c}{2}(t - \sin t), \qquad y = \frac{c}{2}(1 - \cos t).$$

(Take $(x_1, y_1) = (0, 0)$.)

**1-13**   (a)   Prove that the absolute minimum of the functional

$$J(x) = \int_{t_0}^{t_1} p^2(t)\dot{x}^2 \, dt, \qquad x(t_0) = x_0, \quad x(t_1) = x_1 \qquad (*)$$

is given by

$$x(t) = \frac{(x_1 - x_0)^2}{\int_{t_0}^{t_1} (1/p(t))^2 \, dt}$$

and that this minimum is attained if and only if $\dot{x} = C/p(t)^2$, where $C$ is an arbitrary constant.

[*Hint*: Use the Cauchy–Schwarz inequality

$$\left( \int_{t_0}^{t_1} xy \, dt \right) \le \int_{t_0}^{t_1} x^2 \, dt \cdot \int_{t_0}^{t_1} y^2 \, dt.]$$

(b)   Show that $\dot{x}p^2 = C$ is a first integral of the Euler–Lagrange equation corresponding to the functional $(*)$ above.

**1-14**   Prove that the shortest distance between two points in the plane is a straight line.

# CHAPTER 2

# Invariance of Single Integrals

## 2.1  *r*-PARAMETER TRANSFORMATIONS

In this chapter we shall discuss the invariance properties of single integrals under transformations which depend upon *r* parameters. If the fundamental integral is invariant under such transformations, then certain identities involving the Lagrangian can be written down explicitly. To motivate and introduce this idea, we consider a classical result concerning an integration of the Euler–Lagrange equations. By Remark 1.2(b) we know that if the Lagrangian *L* does not depend explicitly on *t*, i.e., the functional is of the form

$$J(x) = \int_{t_0}^{t_1} L(x(t), \dot{x}(t))\, dt,$$

then the Euler–Lagrange equations have first integral

$$L - \dot{x}^k \frac{\partial L}{\partial \dot{x}^k} = C,$$

where *C* is a constant. To say that *L* does not depend explicitly on *t* could be reinterpreted as to say that the functional *J* is invariant under the transformation which takes $t$ to $t + \varepsilon$. The Noether theorem, which is one of the fundamental results of this chapter, can to a certain degree be considered a generalization of the above fact. It states that there is, in general, a connection between the existence of first integrals of the Euler–Lagrange equations and the invariance of the fundamental integral.

To be quite specific about our assumptions, let us consider a Lagrange function $L: I \times R^n \times R^n \to R^1$, where $I \subseteq R^1$ is an open interval of real numbers, and let us assume that $L$ is of class $C^2$ in each of its $2n + 1$ arguments. We then form the variation integral

$$J(x) = \int_a^b L(t, x(t), \dot{x}(t))\, dt, \tag{2.1}$$

where $[a, b] \subseteq I$. As in Chapter 1, we assume that $x \in C_n^2[a, b]$.

The types of invariance transformations that will be considered are transformations of configuration space, i.e., $(t, x^1, \ldots, x^n)$-space, which depend upon $r$ real, independent parameters $\varepsilon^1, \ldots, \varepsilon^r$. In Noether's original paper [1, 2], as well as in more recent treatments of invariance problems, it is assumed that the transformations form a group; in fact, the groups considered are precisely local Lie groups of transformations. In the present work, however, we shall make less stringent assumptions on the transformations. Although many of the transformations which arise in applications actually are groups of transformations (e.g., the Galilean group, the Lorentz group, or the special conformal group), the group concept is not required for the investigation of invariant variation problems. To be more precise, we require here that the transformations are given by

$$\begin{aligned} \bar{t} &= \phi(t, x, \varepsilon) \\ \bar{x}^k &= \psi^k(t, x, \varepsilon) \qquad (k = 1, \ldots, n), \end{aligned} \tag{2.2}$$

where $\varepsilon = (\varepsilon^1, \ldots, \varepsilon^r)$. It is assumed that a domain $U$ in $R^r$ is given which contains the origin as an interior point and that the mappings

$$\begin{aligned} \phi &: [a, b] \times R^n \times U \to I, \\ \psi^k &: [a, b] \times R^n \times U \to R^1 \qquad (k = 1, \ldots, n) \end{aligned}$$

are of class $C^2$ in each of their $1 + n + r$ arguments with

$$\begin{aligned} \phi(t, x, 0) &= t, \\ \psi^k(t, x, 0) &= x^k \qquad (k = 1, \ldots, n). \end{aligned} \tag{2.3}$$

In passing, we note that the right-hand sides of (2.2) can be expanded in a Taylor series about $\varepsilon = 0$ to obtain

$$\begin{aligned} \bar{t} &= t + \tau_s(t, x)\varepsilon^s + o(\varepsilon) \\ \bar{x}^k &= x^k + \xi_s^{\,k}(t, x)\varepsilon^s + o(\varepsilon), \end{aligned} \tag{2.4}$$

where $s$ ranges over $1, \ldots, r$ (summed!) and $o(\varepsilon)$ denote terms which go to zero faster than $|\varepsilon| = (\sum(\varepsilon^s)^2)^{1/2}$; i.e.,

$$\lim_{|\varepsilon| \to 0} \frac{o(\varepsilon)}{|\varepsilon|} = 0.$$

The principal linear parts $\tau_s$ and $\xi_s{}^k$ of $\bar{t}$ and $\bar{x}^k$ with respect to $\varepsilon$ at $\varepsilon = 0$ are commonly called the *infinitesimal generators* of the transformations $\phi$ and $\psi^k$; they are given by

$$\tau_s(t, x) = \frac{\partial \phi}{\partial \varepsilon^s}(t, x, 0), \qquad \xi_s{}^k(t, x) = \frac{\partial \psi^k}{\partial \varepsilon^s}(t, x, 0). \tag{2.5}$$

The transformation defined by (2.4) is classically known as the infinitesimal transformation associated with (2.2).

**2.1  Example**  A one-parameter transformation of the $(t, x)$-plane is given by

$$\bar{t} = t \cos \varepsilon - x \sin \varepsilon, \qquad \bar{x} = t \sin \varepsilon + x \cos \varepsilon.$$

Geometrically, it is a rotation through an angle $\varepsilon$. By expanding $\sin \varepsilon$ and $\cos \varepsilon$ about $\varepsilon = 0$ we obtain the "infinitesimal" rotation

$$\bar{t} = t - \varepsilon x + o(\varepsilon), \qquad \bar{x} = x + \varepsilon t + o(\varepsilon).$$

In this case, the generators are given by $\tau = -x$ and $\xi = t$.

Before defining what is meant by saying that the fundamental integral (2.1) is invariant under the transformations (2.2), there is one additional preliminary notion to discuss, namely the action of the transformations (2.2) on a curve in configuration space. To this end, let $x: [a, b] \to R^n$ be an arbitrary curve of class $C_n{}^2$ given by $x = x(t)$. We shall demonstrate that for sufficiently small $\varepsilon$, the transformations (2.2) carry this curve into an $r$-parameter family of curves $\bar{x} = \bar{x}(\bar{t})$ in $(\bar{t}, \bar{x})$-space when we subject the curve $x = x(t)$ to (2.2) via

$$\bar{t} = \phi(t, x(t), \varepsilon), \qquad \bar{x}^k = \psi^k(t, x(t), \varepsilon).$$

The major step in this construction is contained in the following lemma.

**2.1  Lemma**  Let $x: [a, b] \to R^n$ be a curve of class $C_n{}^2$, then there exists a $d > 0$, depending upon $x(t)$, such that whenever $\varepsilon \in U$ and $|\varepsilon| < d$ the transformation

$$\bar{t} = \phi(t, x(t), \varepsilon), \tag{2.6}$$

where $\phi$ is given by (2.2), has a unique inverse.

*Proof*   Define the $r$-parameter family of functions

$$\gamma\colon [a, b] \times U \to I$$

by

$$\gamma(t, \varepsilon) = \phi(t, x(t), \varepsilon).$$

From (2.3) it follows that

$$\gamma(t, 0) = t$$

and so

$$\frac{\partial\gamma}{\partial t}(t_0, 0) = 1$$

for each $t_0 \in [a, b]$. Since $\partial\gamma/\partial t$ is continuous at $(t_0, 0)$, there is a $\delta(t_0) > 0$ and a $d(t_0) > 0$ for which

$$\frac{\partial\gamma}{\partial t}(t, \varepsilon) > 0$$

for all $t \in (t_0 - \delta(t_0), t_0 + \delta(t_0)) \cap [a, b]$ and for all $\varepsilon \in U$ with $|\varepsilon| < d(t_0)$. We note that the intervals $(t_0 - \delta(t_0), t_0 + \delta(t_0))$ cover $[a, b]$, and since $[a, b]$ is compact, there exists a finite collection of points $\{t_1, \ldots, t_m\} \subset [a, b]$ such that

$$\bigcup_{i=1}^{m} (t_i - \delta(t_i), t_i + \delta(t_i))$$

cover $[a, b]$. Setting $d = \min\{d(t_1), \ldots, d(t_m)\}$, we observe that for any $t \in [a, b]$ and $|\varepsilon| < d$, $\varepsilon \in U$ we have $(\partial\gamma/\partial t)(t, \varepsilon) > 0$. Consequently, the mapping $\gamma_\varepsilon\colon [a, b] \to I$ defined by $\gamma_\varepsilon(t) = \gamma(t, \varepsilon)$ for $|\varepsilon| < d$, $\varepsilon \in U$, is monotonically increasing and therefore invertible. This completes the proof.   $\square$

Now, for sufficiently small $\varepsilon$ we may define the transformed curves $\bar{x} = \bar{x}(\bar{t})$ in $(\bar{t}, \bar{x})$-space. Specifically, since $\bar{t} = \phi(t, x(t), \varepsilon)$ is invertible we may solve it for $t$ in terms of $\bar{t}$ to obtain

$$t = T(\bar{t}, \varepsilon).$$

Substitution into the equation

$$\bar{x}^k = \psi^k(t, x(t), \varepsilon)$$

yields

$$\begin{aligned}\bar{x}^k &= \psi^k(T(\bar{t}, \varepsilon), x(T(\bar{t}, \varepsilon)), \varepsilon) \\ &\equiv \bar{x}^k(\bar{t}).\end{aligned} \tag{2.7}$$

Hence, from (2.7), the defining relation for $\bar{x} = \bar{x}(\bar{t})$ may be written as

$$\bar{x}^k(\phi(t, x(t), \varepsilon)) = \psi^k(t, x(t), \varepsilon). \tag{2.8}$$

Clearly, $\bar{x}(\bar{t})$ is defined on $[\bar{a}, \bar{b}]$, where

$$\bar{a} = \phi(a, x(a), \varepsilon), \qquad \bar{b} = \phi(b, x(b), \varepsilon). \tag{2.9}$$

To aid in the understanding of the above formulation, we can realize a geometrical interpretation in the special case that $n = r = 1$. Here, $x = x(t)$ is a function in the $tx$-plane. Under the transformation $t \to \bar{t} = \phi(t, x(t), \varepsilon)$, $x \to \bar{x} = \psi(t, x(t), \varepsilon)$, the curve $x = x(t)$ gets mapped to a one-parameter family of curves, whose graphs are the graphs of functions in the $\bar{t}\bar{x}$-plane, provided that $\varepsilon$ is sufficiently small. Figure 3 depicts this transformation; the curve drawn in the $\bar{t}\bar{x}$-plane shows one curve out of the one-parameter family; i.e., it shows the transformation for one particular value of $\varepsilon$.

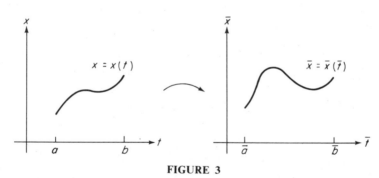

**FIGURE 3**

**2.2  Example**  In the $tx$-plane consider the curve $x(t) = mt$, where $m$ is constant. Then under the infinitesimal rotation (see Example 2.1)

$$\bar{t} = t - \varepsilon x,$$
$$\bar{x} = x + \varepsilon t,$$

we have

$$\bar{t} = t - \varepsilon mt,$$
$$\bar{x} = mt + \varepsilon t.$$

Solving the first equation for $t$ in terms of $\bar{t}$, we obtain

$$t = \frac{\bar{t}}{1 - \varepsilon m}.$$

Thus

$$\bar{x} = \frac{m + \varepsilon}{1 - m\varepsilon} \, \bar{t} \equiv \bar{x}(\bar{t}).$$

Therefore, a straight line through the origin with slope $m$ gets mapped to a one-parameter family of straight lines in the $tx$-plane with slopes $(m + \varepsilon)/(1 - m\varepsilon)$.

## 2.2  INVARIANCE DEFINITIONS

We are now in position to define invariance of the variation integral. By way of motivation, consider Fig. 3 when $n = r = 1$. In the $tx$-plane we can calculate the functional

$$J(x) = \int_a^b L(t, x(t), \dot{x}(t)) \, dt.$$

In the $\bar{t}\bar{x}$-plane we are able to calculate

$$J(\bar{x}) = \int_{\bar{a}}^{\bar{b}} L\left(\bar{t}, \bar{x}(\bar{t}), \frac{d\bar{x}}{d\bar{t}}(\bar{t})\right) d\bar{t}.$$

Clearly, we would like to say that $J$ is invariant under the given transformation if $J(x) = J(\bar{x})$, or at least so up to first-order terms in $\varepsilon$. A precise definition is as follows:

**2.1  Definition**  The fundamental integral (2.1) is absolutely invariant under the $r$-parameter family of transformations (2.2) if and only if given any curve $x: [a, b] \to R^n$ of class $C_n{}^2$ and $a \le t_1 < t_2 \le b$ we have

$$\int_{\bar{t}_1}^{\bar{t}_2} L\left(\bar{t}, \bar{x}(\bar{t}), \frac{d\bar{x}(\bar{t})}{d\bar{t}}\right) d\bar{t} - \int_{t_1}^{t_2} L(t, x(t), \dot{x}(t)) \, dt = o(\varepsilon) \qquad (2.10)$$

for all $\varepsilon \in U$ with $|\varepsilon| < d$, where $d$ depends upon $x(t)$, where $\bar{x}(\bar{t})$ is defined by (2.8), and where $\bar{t}_i = \phi(t_i, x(t_i), \varepsilon)$, $i = 1, 2$. Therefore, the fundamental integral is invariant under (2.2) if the difference between the transformed integral and the nontransformed integral is equal to first-order terms in $\varepsilon$.

**2.1  Remark**  It is clear, by transforming the first integral in (2.10) back to the interval $[t_1, t_2]$, that condition (2.10) is entirely equivalent to the condition

$$L\left(\bar{t}, \bar{x}(\bar{t}), \frac{d\bar{x}(\bar{t})}{d\bar{t}}\right) \frac{d\bar{t}}{dt} - L(t, x(t), \dot{x}(t)) = o(\varepsilon), \qquad (2.11)$$

which involves only the Lagrangian and not the action integral.

In order to determine whether or not a given integral is invariant under a given transformation, it suffices to directly determine the validity of (2.10) or (2.11) for the particular problem, as the following example shows.

**2.3   Example**   We consider the arclength functional

$$J(x) = \int_a^b \sqrt{1 + \dot{x}(t)^2}\, dt \tag{2.12}$$

and the one-parameter family of transformations

$$\bar{t} = t - \varepsilon x, \qquad \bar{x} = x + \varepsilon t, \tag{2.13}$$

where $\varepsilon$ is the parameter. Geometrically, since (2.13) represents an "infinitesimal" rotation of axes, it is obvious that the arclength functional (2.12) is invariant under (2.13). However, a direct calculation of this invariance will illustrate the general procedure for determining the invariance properties of any given functional. We note that, from (2.13),

$$\frac{d\bar{t}}{dt} = 1 - \varepsilon\dot{x}, \qquad \frac{d\bar{x}}{dt} = \dot{x} + \varepsilon,$$

and consequently,

$$\frac{d\bar{x}}{d\bar{t}} = \frac{d\bar{x}}{dt}\frac{dt}{d\bar{t}}$$

$$= \frac{d\bar{x}}{dt}\bigg/\frac{d\bar{t}}{dt}$$

$$= \frac{\dot{x} + \varepsilon}{1 - \varepsilon\dot{x}}.$$

Therefore, condition (2.11) becomes

$$\sqrt{1 + \left(\frac{d\bar{x}}{d\bar{t}}\right)^2}\frac{d\bar{t}}{dt} - \sqrt{1 + \dot{x}^2} = \sqrt{1 + \left(\frac{\dot{x} + \varepsilon}{1 - \varepsilon\dot{x}}\right)^2}(1 - \varepsilon\dot{x}) - \sqrt{1 + \dot{x}^2}$$

$$= \sqrt{1 + \dot{x}^2}(\sqrt{1 + \varepsilon^2} - 1)$$

$$= \sqrt{1 + \dot{x}^2}(1 + \tfrac{1}{2}\varepsilon^2 + \tfrac{1}{8}\varepsilon^4 + \cdots - 1)$$

$$= o(\varepsilon).$$

Hence, according to Definition 2.1, the functional $J$ defined by (2.12) is absolutely invariant under the transformations (2.13).

For some applications we shall require a more general definition of invariance than the one given above. This need will arise, for example, in determining the invariance properties of the action integral for the $n$-body problem. Briefly, the generalization consists of replacing the $o(\varepsilon)$ on the right-hand side of (2.11) by a term which is linear in the $\varepsilon^s$ with the coefficients being total derivatives. We shall say that the fundamental integral is *divergence-invariant*, or invariant up to a divergence term, if there exist $r$ functions $\Phi_s: I \times R^n \to R^1$, $s = 1, \ldots, r$, of class $C^1$ such that

$$L\left(\bar{t}, \bar{x}(\bar{t}), \frac{d\bar{x}(\bar{t})}{d\bar{t}}\right)\frac{d\bar{t}}{dt} - L(t, x(t), \dot{x}(t)) = \varepsilon^s \frac{d\Phi_s}{dt}(t, x(t)) + o(\varepsilon), \qquad (2.14)$$

with the remaining conditions of Definition 2.1 holding true.


## 2.3   THE FUNDAMENTAL INVARIANCE IDENTITIES

Under the assumption that the fundamental integral is invariant (either absolutely or up to a divergence) under an $r$-parameter family of transformations, we shall now derive a set of basic invariance identities which involve the Lagrangian, its derivatives, and the infinitesimal generators of the transformations. The fundamental theorem is the following:

**2.1   Theorem**   A necessary condition for the fundamental integral (2.1) to be divergence-invariant under the $r$-parameter transformations (2.2) is that the Lagrangian $L(t, x, \dot{x})$ and its derivatives satisfy the $r$ identities

$$\frac{\partial L}{\partial t}\tau_s + \frac{\partial L}{\partial x^k}\xi_s^k + \frac{\partial L}{\partial \dot{x}^k}\left(\frac{d\xi_s^k}{dt} - \dot{x}^k\frac{d\tau_s}{dt}\right) + L\frac{d\tau_s}{dt} = \frac{d\Phi_s}{dt} \qquad (2.15)$$

$(s = 1, \ldots, r)$, where the $\tau_s$ and $\xi_s^k$ are defined by the transformations according to (2.5).

*Proof*   In the proof we shall need the following expressions which are easily derived from (2.2):

$$\left(\frac{\partial \bar{t}}{\partial t}\right)_0 = 1, \qquad \left(\frac{\partial \bar{x}^k}{\partial t}\right)_0 = \left(\frac{\partial \bar{t}}{\partial x^k}\right)_0 = 0, \qquad \left(\frac{\partial \bar{x}^k}{\partial x^h}\right)_0 = \delta_h^k,$$

$$\left(\frac{\partial^2 \bar{x}^k}{\partial \varepsilon^s\, \partial t}\right)_0 = \frac{\partial \xi_s^k}{\partial t}, \qquad \left(\frac{\partial^2 \bar{x}^k}{\partial \varepsilon^s\, \partial x^h}\right)_0 = \frac{\partial \xi_s^k}{\partial x^h}, \qquad (2.16)$$

$$\left(\frac{\partial^2 \bar{t}}{\partial \varepsilon^s\, \partial t}\right)_0 = \frac{\partial \tau_s}{\partial t}, \qquad \left(\frac{\partial^2 \bar{t}}{\partial \varepsilon^s\, \partial x^h}\right)_0 = \frac{\partial \tau_s}{\partial x^h},$$

where $\delta_h{}^k$ is the Kronecker delta symbol, and $(\cdot)_0$ denotes $(\cdot)_{\varepsilon=0}$. The basic idea is to differentiate (2.14) with respect to $\varepsilon^s$ and then set $\varepsilon = 0$. Performing this calculation we obtain

$$\left(\frac{d\bar{t}}{dt}\right)_0\left(\frac{\partial L}{\partial t}\,\tau_s + \frac{\partial L}{\partial x^k}\,\xi_s{}^k + \frac{\partial L}{\partial \dot{x}^k}\left(\frac{\partial}{\partial \varepsilon^s}\frac{d\bar{x}^k}{d\bar{t}}\right)_0\right) + L\left(\frac{\partial}{\partial \varepsilon^s}\frac{d\bar{t}}{dt}\right)_0 = \frac{d\Phi_s}{dt}. \quad (2.17)$$

We now calculate the terms

$$\left(\frac{\partial}{\partial \varepsilon^s}\frac{d\bar{t}}{dt}\right)_0 \qquad \text{and} \qquad \left(\frac{\partial}{\partial \varepsilon^s}\frac{d\bar{x}^k}{d\bar{t}}\right)_0.$$

First, we note that by the chain rule,

$$\frac{d\bar{t}}{dt} = \frac{\partial \bar{t}}{\partial t} + \frac{\partial \bar{t}}{\partial x^k}\,\dot{x}^k.$$

Differentiating this equation with respect to $\varepsilon^s$, setting $\varepsilon = 0$, and finally using (2.16), we get

$$\left(\frac{\partial}{\partial \varepsilon^s}\frac{d\bar{t}}{dt}\right)_0 = \frac{\partial \tau_s}{\partial t} + \frac{\partial \tau_s}{\partial x^k}\,\dot{x}^k = \frac{d\tau_s}{dt}. \quad (2.18)$$

To calculate the other expression in (2.17), we differentiate (2.7) with respect to $t$ to obtain

$$\frac{\partial \bar{x}^k}{\partial t} + \frac{\partial \bar{x}^k}{\partial x^h}\,\dot{x}^h = \frac{d\bar{x}^k}{d\bar{t}}\frac{d\bar{t}}{dt} = \frac{d\bar{x}^k}{d\bar{t}}\left(\frac{\partial \bar{t}}{\partial t} + \frac{\partial \bar{t}}{\partial x^h}\,\dot{x}^h\right). \quad (2.19)$$

Setting $\varepsilon = 0$ in this expression and using (2.16), we conclude that

$$\delta_h{}^k\dot{x}^h = \left(\frac{d\bar{x}^k}{d\bar{t}}\right)_0(1 + 0\cdot x^h)$$

or

$$\dot{x}^k = \left(\frac{d\bar{x}^k}{d\bar{t}}\right)_0.$$

Now, differentiating (2.19) with respect to $\varepsilon^s$ we obtain

$$\frac{\partial^2 \bar{x}^k}{\partial \varepsilon^s\,\partial t} + \frac{\partial^2 \bar{x}^k}{\partial \varepsilon^s\,\partial x^h}\,\dot{x}^h = \frac{d\bar{x}^k}{d\bar{t}}\left(\frac{\partial^2 \bar{t}}{\partial \varepsilon^s\,\partial t} + \frac{\partial^2 \bar{t}}{\partial \varepsilon^s\,\partial x^h}\,\dot{x}^h\right) + \frac{\partial}{\partial \varepsilon^s}\frac{d\bar{x}^k}{d\bar{t}}\left(\frac{\partial \bar{t}}{\partial t} + \frac{\partial \bar{t}}{\partial x^h}\,\dot{x}^h\right),$$

after which we set $\varepsilon = 0$ while using (2.16). This yields

$$\frac{\partial \xi_s{}^k}{\partial t} + \frac{\partial \xi_s{}^k}{\partial x^h}\,\dot{x}^h = \dot{x}^k\left(\frac{\partial \tau_s}{\partial t} + \frac{\partial \tau_s}{\partial x^h}\,\dot{x}^h\right) + \left(\frac{\partial}{\partial \varepsilon^s}\frac{d\bar{x}^k}{d\bar{t}}\right)_0,$$

which in turn gives

$$\left(\frac{\partial}{\partial \varepsilon^s} \frac{d\bar{x}^k}{d\bar{t}}\right)_0 = \frac{d\xi_s^{\ k}}{dt} - \dot{x}^k \frac{d\tau_s}{dt}.$$

Substituting this and (2.18) into (2.17) and observing that $(d\bar{t}/dt)_{\varepsilon=0} = 1$, we obtain the identities (2.15) and the proof is complete. $\square$

**2.2 Remark** The fundamental invariance identities (2.15) express a condition involving the Lagrange function $L$ and the infinitesimal generators $\tau_s$ and $\xi_s^{\ k}$ of the transformation. We note that the arguments $t$ and $x$ which occur in (2.15) refer to the arbitrary curve $x = x(t)$, and hence (2.15) represent identities in $t$, $x^k$, and $\dot{x}^k$ for arbitrary directional arguments $\dot{x}^k$.

**2.3 Remark** If the Lagrangian and the transformation are known, then Eqs. (2.15) can be written down directly to obtain some fundamental relations for the system. As we shall see in the next section, these relations lead to conservation laws for the system. Just as important, however, is the following interpretation. If only the transformation is known, Eqs. (2.15) represent $r$ first-order quasi-linear partial differential equations in the unknown Lagrange function $L$. Therefore this interpretation leads to a method of characterizing the set of all Lagrangians which possess given invariance properties. Physically, for example, this permits one to determine classes variational problems which are Galilean or Lorentz invariant, say. We shall discuss this aspect of the subject in subsequent sections.

## 2.4  THE NOETHER THEOREM AND CONSERVATION LAWS

Now we shall show that the classical theorem of Emmy Noether on invariant variational problems can be derived as a corollary of Theorem (2.1). As in Chapter 1, let us denote the *Euler–Lagrange expressions* by $E_k$; i.e.,

$$E_k \equiv \frac{\partial L}{\partial x^k} - \frac{d}{dt} \frac{\partial L}{\partial \dot{x}^k} \qquad (k = 1, \ldots, n). \tag{2.20}$$

Then we have:

**2.2 Theorem** (*Noether*) Under the hypotheses of Theorem (2.1), there exist $r$ identities of the form

$$-E_k(\xi_s^{\ k} - \dot{x}^k \tau_s) = \frac{d}{dt}\left(\left(L - \dot{x}^k \frac{\partial L}{\partial \dot{x}^k}\right)\tau_s + \frac{\partial L}{\partial \dot{x}^k}\xi_s^{\ k} - \Phi_s\right) \tag{2.21}$$

$(s = 1, \ldots, r)$, where $\tau_s$ and $\xi_s^{\ k}$ are given by (2.5), and $\Phi_s$ is defined in (2.14).

**2.4 Remark** This theorem is oftentimes expressed in words by saying that under the invariance hypothesis there exists $r$ linear combinations of the Euler–Lagrange expressions $E_k$ which are exact differentials. Identities (2.21) are known as the *Noether identities*; we note that they inherently involve second derivatives of the functions $x^k(t)$ through the presence of the expressions $E_k$. This is in contrast to the identities (2.15) which involve only first derivatives of the $x^k(t)$.

*Proof of Theorem* 2.2 To derive (2.21) from (2.15) we must introduce the expressions $E_k$ into (2.15). To this end, observe that the following identities are valid:

$$\frac{\partial L}{\partial \dot{x}^k}\frac{d\xi_s^{\ k}}{dt} = \frac{d}{dt}\left(\frac{\partial L}{\partial \dot{x}^k}\xi_s^{\ k}\right) - \frac{d}{dt}\frac{\partial L}{\partial \dot{x}^k}\xi_s^{\ k},$$

$$\frac{\partial L}{\partial t}\tau_s = \frac{dL}{dt}\tau_s - \frac{\partial L}{\partial x^k}\dot{x}^k\tau_s - \frac{\partial L}{\partial \dot{x}^k}\ddot{x}^k\tau_s,$$

$$\frac{\partial L}{\partial \dot{x}^k}\dot{x}^k\frac{d\tau_s}{dt} + \frac{\partial L}{\partial \dot{x}^k}\ddot{x}^k\tau_s = \frac{d}{dt}\left(\frac{\partial L}{\partial \dot{x}^k}\dot{x}^k\tau_s\right) - \frac{d}{dt}\left(\frac{\partial L}{\partial \dot{x}^k}\right)\dot{x}^k\tau_s.$$

Substituting these into (2.15) yields, after simplification, the identities (2.21). Verification is left for the reader. $\square$

The importance of Theorem 2.2 lies in the following trivial consequence.

**2.3 Theorem** If the fundamental integral (2.1) is divergence-invariant under the $r$-parameter group of transformations (2.2), and if $E_k = 0$ for $k = 1, \ldots, n$, then the following $r$ expressions hold true:

$$\Psi_s \equiv \left(L - \dot{x}^k\frac{\partial L}{\partial \dot{x}^k}\right)\tau_s + \frac{\partial L}{\partial \dot{x}^k}\xi_s^{\ k} - \Phi_s = \text{constant} \qquad (s = 1, \ldots, r). \quad (2.22)$$

**2.5 Remark** Since the expressions $\Psi_s$ defined in (2.22) are constant whenever $E_k = 0\,(k = 1, \ldots, n)$, they are *first integrals* of the differential equations of motion, i.e., the Euler–Lagrange equations. In physical applications, Eqs. (2.22) are thus interpreted as conservation laws of the system whose governing equations are $E_k = 0$. In still other words, the expressions $\Psi_s$ are constant along any extremal.

**2.6 Remark** If we define the *canonical momenta* by

$$p_k \equiv \frac{\partial L}{\partial \dot{x}^k},$$

and the *Hamiltonian*, which in some mechanical systems is the total energy, by

$$H \equiv -L + \dot{x}^k\frac{\partial L}{\partial \dot{x}^k},$$

then Eqs. (2.22) may be written simply in terms of these canonical variables as

$$-H\tau_s + p_k \xi_s^k - \Phi_s = \text{constant} \qquad (s = 1, \ldots, r). \qquad (2.23)$$

The classical result that we mentioned in the introduction in Section 2.1 concerning an integration of the Euler–Lagrange equations follows easily from Theorem 2.3. If the fundamental integral (2.1) is invariant under the one-parameter transformation

$$\bar{t} = t + \varepsilon, \qquad \bar{x}^k = x^k, \qquad (2.24)$$

then $\tau_s = 1$, $\xi_s^k = 0$ and $\Phi_s = 0$. Consequently, a first integral is given according to (2.22) or (2.23) by

$$-H = L - \dot{x}^k \frac{\partial L}{\partial \dot{x}^k} = \text{constant}. \qquad (2.25)$$

With respect to dynamical systems, we can interpret transformation (2.24) as a time translation. Consequently, Theorem 2.3 shows that if the fundamental integral is invariant under a time translation then the Hamiltonian, or total energy of the system, is constant. We obtain therefore the conservation of energy theorem.

As another example, suppose the integral (2.1) is absolutely invariant under translations of the form

$$\bar{t} = t, \qquad \bar{x}^k = x^k + \varepsilon^k,$$

which define an $n$-parameter family of transformations with generators $\tau_s = 0$ and $\xi_s^k = \delta_s^k$; in this case, the first integrals are given from (2.22) by

$$\frac{\partial L}{\partial \dot{x}^k} \delta_s^k = \text{constant}$$

or

$$p_s = \frac{\partial L}{\partial \dot{x}^s} = \text{constant} \qquad (s = 1, \ldots, n).$$

For dynamical systems, these equations can be interpreted as a statement of conservation of linear momentum.

**2.4 Example**   A particle of mass $m$ moves in a plane and is attracted to the origin by a force inversely proportional to the square of its distance from the origin. In this case the action integral is

$$J(r, \theta) = \int_{t_0}^{t_1} \left\{ \frac{1}{2} m(\dot{r}^2 + r^2 \dot{\theta}^2) + \frac{A}{r} \right\} dt,$$

where $A$ is constant and $r$ and $\theta$ are the polar coordinates of the particle. We consider the one-parameter family of rotations (here, of course, $x^1 = r$, $x^2 = \theta$)

$$\bar{t} = t, \qquad \bar{r} = r, \qquad \bar{\theta} = \theta + \varepsilon,$$

and since $d\theta/d\bar{t} = d\theta/dt$, it follows that the action integral is absolutely invariant under this family. Then Noether's theorem in this context states that

$$\left( L - \dot{r}\frac{\partial L}{\partial \dot{r}} - \dot{\theta}\frac{\partial L}{\partial \dot{\theta}} \right)\tau + \frac{\partial L}{\partial \dot{r}}\rho + \frac{\partial L}{\partial \dot{\theta}}\sigma = \text{constant},$$

where $\tau$, $\rho$, and $\sigma$ are the generators given by

$$\tau \equiv \left.\frac{\partial \bar{t}}{\partial \varepsilon}\right|_0 = 0, \qquad \rho \equiv \left.\frac{\partial \bar{r}}{\partial \varepsilon}\right|_0 = 0, \qquad \sigma \equiv \left.\frac{\partial \bar{\theta}}{\partial \varepsilon}\right|_0 = 1.$$

Consequently,

$$\frac{\partial L}{\partial \dot{\theta}} = \text{constant} \qquad \text{or} \qquad mr^2\dot{\theta} = \text{constant}.$$

Geometrically, this first integral of the motion states that equal areas are swept out in equal times.

Since the action integral is also explicitly independent of $t$, it follows from (2.25) that

$$-H = L - \dot{\theta}L_{\dot{\theta}} - \dot{r}L_r = \text{constant}$$

or

$$\tfrac{1}{2}m(\dot{r}^2 + r^2\dot{\theta}^2) - \frac{A}{r} = \text{constant}.$$

This is a statement of conservation of energy for the system.

## 2.5   PARTICLE MECHANICS AND THE GALILEAN GROUP

In this section we give a fairly detailed treatment of the application of the Noether theorem to compute the conservation laws in particle mechanics. This application was first carried out by Bessel-Hagen [1] in 1921 using the ten-parameter Galilean group consisting of a time translation, spatial translations and rotations, and velocity translations. The resulting ten conservation laws are the classical ten integrals of the $n$-body problem: conservation of energy, linear momentum, angular momentum, and the uniform motion of the center of mass.

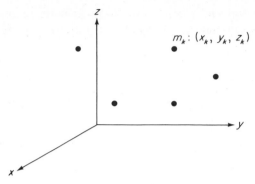

**FIGURE 4**

We consider a system of $n$ particles, the $k$th particle having mass $m_k$ and position $(x_k, y_k, z_k)$ in the usual rectangle coordinate system of $R^3$ (see Fig. 4). We assume in general that the potential energy of the system is a function

$$V = V(t, x_1, \ldots, x_n, y_1, \ldots, y_n, z_1, \ldots, z) \qquad (2.26)$$

of $3n + 1$ variables, which we denote simply by $V(t, x, y, z)$. The kinetic energy of the $k$th particle is $\frac{1}{2}m_k(\dot{x}_k^2 + \dot{y}_k^2 + \dot{z}_k^2)$, and the total kinetic energy $T$ of the system is sum of the kinetic energies of each particle or

$$T = \frac{1}{2}\sum_k m_k(\dot{x}_k^2 + \dot{y}_k^2 + \dot{z}_k^2) \qquad (2.27)$$

(in this section, $\sum_k$ will mean $\sum_{k=1}^n$). The Lagrangian is then given by the difference between the kinetic and potential energies, or

$$L = \frac{1}{2}\sum_k m_k(\dot{x}_k^2 + \dot{y}_k^2 + \dot{z}_k^2) - V(t, x, y, z), \qquad (2.28)$$

and Hamilton's principle assures us that the motion of the system takes place in such a manner that the functional $J = \int L\, dt$ is stationary. A necessary condition for $J$ to be extremal is that the time-dependent coordinates $x_1, \ldots, x_n, y_1, \ldots, y_n, z_1, \ldots, z_n$ satisfy the system of $3n$ Euler–Lagrange equations, which in this case coincide with Newton's equations of motion:

$$-\frac{\partial V}{\partial x_k} = m_k \ddot{x}_k, \qquad -\frac{\partial V}{\partial y_k} = m_k \ddot{y}_k, \qquad -\frac{\partial V}{\partial z_k} = m_k \ddot{z}_k. \qquad (2.29)$$

It is of course possible to deduce the conservation theorems directly from the equations of motion (2.29). Our purpose here, however, is to show that these first integrals follow from the fact that the fundamental integral $\int L\, dt$ is Galilean invariant. To this end, we now define the Galilean group, which is a ten-parameter local Lie group and of the type considered in Section 2.2; the independent parameters are denoted by $\varepsilon^1, \ldots, \varepsilon^{10}$.

(i)   *Time translation*

$$\bar{t} = t + \varepsilon^1,$$
$$\bar{x}_k = x_k, \qquad \bar{y}_k = y_k, \qquad \bar{z}_k = z_k. \tag{2.30}$$

(ii)   *Spatial translations*

$$\bar{t} = t,$$
$$\bar{x}_k = x_k + \varepsilon^2, \qquad \bar{y}_k = y_k + \varepsilon^3, \qquad \bar{z}_k = z_k + \varepsilon^4. \tag{2.31}$$

(iii)   *Spatial rotations*

$$\bar{t} = t,$$
$$\bar{x}_k = x_k + \varepsilon^5 y_k + \varepsilon^6 z_k,$$
$$\bar{y}_k = y_k - \varepsilon^5 x_k + \varepsilon^7 z_k, \tag{2.32}$$
$$\bar{z}_k = z_k - \varepsilon^6 x_k - \varepsilon^7 y_k.$$

(iv)   *Velocity translations*

$$\bar{t} = t,$$
$$\bar{x}_k = x_k + \varepsilon^8 t, \qquad \bar{y}_k = y_k + \varepsilon^9 t, \qquad \bar{z}_k = z_k + \varepsilon^{10} t. \tag{2.33}$$

In the sequel, we shall denote the generators of the transformations $t \to \bar{t}$, $x \to \bar{x}$, $y \to \bar{y}$, and $z \to \bar{z}$ by $\tau$, $\xi$, $\eta$ and $\zeta$, respectively.

In the case of time translations (2.30), we make the further supposition that the potential energy $V$ is not explicitly dependent upon time. Then the remarks at the end of section (2.4) apply; that is to say, the fundamental integral $\int (T - V)\, dt$ is invariant under (2.30), and therefore by (2.25) we have

$$L - \dot{x}_k \frac{\partial L}{\partial \dot{x}_k} - \dot{y}_k \frac{\partial L}{\partial \dot{y}_k} - \dot{z}_k \frac{\partial L}{\partial \dot{z}_k} = \text{constant.} \tag{2.34}$$

Clearly,

$$\frac{\partial L}{\partial \dot{x}_k} = m_k \dot{x}_k, \qquad \frac{\partial L}{\partial \dot{y}_k} = m_k \dot{y}_k, \qquad \frac{\partial L}{\partial \dot{z}_k} = m_k \dot{z}_k \qquad \text{(no sum)} \tag{2.35}$$

and thus (2.34) becomes upon substitution of the Lagrangian $L$ given by (2.28), with $V = V(x, y, z)$,

$$\tfrac{1}{2} \sum_k m_k(\dot{x}_k{}^2 + \dot{y}_k{}^2 + \dot{z}_k{}^2) + V(x, y, z) = \text{constant.}$$

This equation expresses the fact that

$$E \equiv T + V = \text{constant},$$

or the total energy $E$ is constant throughout the evolution of the system when $V$ is independent of time. This is the principle of *conservation of energy*.

Let us now assume that the potential energy depends only on the mutual positions of the particles; i.e., $V$ has $3n(n-1)/2$ arguments and is of the form

$$V = V(\dots, x_i - x_j, y_i - y_j, z_i - z_j, \dots) \qquad (1 \le i < j \le n). \quad (2.36)$$

It then follows immediately that $V$ is invariant under the spatial translations (2.31); for,

$$
\begin{aligned}
V(&\dots, \bar{x}_i - \bar{x}_j, \bar{y}_i - \bar{y}_j, \bar{z}_i - \bar{z}_j, \dots) \\
&= V(\dots, (x_i + \varepsilon^2) - (x_j + \varepsilon^2), (y_i + \varepsilon^3) - (y_j + \varepsilon^3), \\
&\qquad (z_i + \varepsilon^4) - (z_j + \varepsilon^4), \dots) \\
&= V(\dots, x_i - x_j, y_i - y_j, z_i - z_j, \dots).
\end{aligned}
$$

To see that the kinetic energy (2.27) is also invariant, it suffices to note that, for example,

$$\frac{d\bar{x}_k}{d\bar{t}} = \frac{d\bar{x}_k}{dt} = \frac{d}{dt}(x_k + \varepsilon^2) = \frac{dx_k}{dt}.$$

Similar relations hold for $y_k$ and $z_k$. Therefore, $\int L \, dt$ is absolutely invariant under spatial translations (2.31), and consequently Theorem 2.3 implies the existence of three conservations laws. In order to deduce these laws, we note from (2.31) that the group generators are given by

$$\tau_s \equiv \frac{\partial \bar{t}}{\partial \varepsilon^s}\bigg|_{\varepsilon=0} = 0 \qquad (s = 2, 3, 4), \tag{2.37}$$

$$\xi_s^{\ k} \equiv \frac{\partial \bar{x}_k}{\partial \varepsilon^s}\bigg|_0 = \begin{cases} 1, & s = 2, \\ 0, & s = 3, 4; \end{cases} \qquad \eta_s^{\ k} \equiv \frac{\partial \bar{y}_k}{\partial \varepsilon^s}\bigg|_0 = \begin{cases} 1, & s = 3, \\ 0, & s = 2, 4; \end{cases}$$

$$\zeta_s^{\ k} \equiv \frac{\partial \bar{z}_k}{\partial \varepsilon^s}\bigg|_0 = \begin{cases} 1, & s = 4, \\ 0, & s = 2, 3. \end{cases}$$

The conservation laws (2.22) then become

$$\sum_k \left( \frac{\partial L}{\partial \dot{x}_k} \xi_s^{\ k} + \frac{\partial L}{\partial \dot{y}_k} \eta_s^{\ k} + \frac{\partial L}{\partial \dot{z}_k} \zeta_s^{\ k} \right) = 0 \qquad (s = 2, 3, 4)$$

or

$$\sum_k m_k \dot{x}_k = \text{constant}, \qquad \sum_k m_k \dot{y}_k = \text{constant}, \qquad \sum_k m_k \dot{z}_k = \text{constant}. \quad (2.38)$$

Therefore, we have shown that the momentum in each coordinate direction remains unchanged if the potential energy is of the form (2.36). Hence, *linear momentum is conserved* if the fundamental integral is invariant under spatial translations. This specific example illustrates the general conservation of momentum statement made at the end of Section (2.4).

*Conservation of angular momentum* follows from the invariance of the fundamental integral under spatial rotations. To obtain the required invariance, let us assume that the potential energy depends only on the mutual distances between the $n$ particles, i.e.,

$$V = V(\ldots, r_{ij}, \ldots), \qquad r_{ij} = ((x_i - x_j)^2 + (y_i - y_j)^2 + (z_i - z_j)^2)^{1/2}$$
$$(1 \le i < j \le n). \quad (2.39)$$

In this case it is obvious that $V(\bar{r}_{ij}) = V(r_{ij})$ since a rotation of axes in $R^3$ is a rigid motion and hence preserves the Euclidean distance between points.

To see that the kinetic energy (2.27) is invariant under rotations, we note, using (2.32), that

$$\frac{d\bar{x}_k}{d\bar{t}} = \frac{d}{dt}(x + \varepsilon^5 y_k + \varepsilon^6 z_k),$$

$$\frac{d\bar{y}_k}{d\bar{t}} = \frac{d}{dt}(y_k - \varepsilon^5 x_k + \varepsilon^7 z_k),$$

$$\frac{d\bar{z}_k}{d\bar{t}} = \frac{d}{dt}(z_k - \varepsilon^6 x_k - \varepsilon^7 y_k).$$

Therefore

$$\left(\frac{d\bar{x}_k}{d\bar{t}}\right)^2 + \left(\frac{d\bar{y}_k}{d\bar{t}}\right)^2 + \left(\frac{d\bar{z}_k}{d\bar{t}}\right)^2 = (\dot{x}_k + \varepsilon^5 \dot{y}_k + \varepsilon^6 \dot{z}_k)^2 + (\dot{y}_k - \varepsilon^5 \dot{x}_k + \varepsilon^7 \dot{z}_k)^2$$
$$+ (\dot{z}_k - \varepsilon^6 \dot{x}_k - \varepsilon^7 \dot{y}_k)^2$$
$$= \dot{x}_k{}^2 + \dot{y}_k{}^2 + \dot{z}_k{}^2 + o(\varepsilon).$$

We have shown, under the assumption (2.39) on the potential energy, that the integral $\int (T - V)\, dt$ is absolutely invariant under spatial rotations. Again, by Theorem 2.3 we know that there exists three first integrals of the governing equations of motion. To determine these integrals we must again write down the generators of the group corresponding to $s = 5, 6, 7$. From (2.32),

$$\tau_s \equiv \left.\frac{\partial \bar{t}}{\partial \varepsilon^s}\right|_0 = 0 \quad (s = 5, 6, 7),$$

$$\xi_s{}^k \equiv \left.\frac{\partial \bar{x}_k}{\partial \varepsilon^s}\right|_0 = \begin{cases} y_k, & s = 5, \\ z_k, & s = 6, \\ 0, & s = 7, \end{cases} \qquad \eta_s{}^k \equiv \left.\frac{\partial \bar{y}_k}{\partial \varepsilon^s}\right|_0 = \begin{cases} -x_k, & s = 5, \\ 0, & s = 6, \quad (2.40) \\ z_k, & s = 7; \end{cases}$$

$$\zeta_s{}^k \equiv \left.\frac{\partial \bar{z}_k}{\partial \varepsilon^s}\right|_0 = \begin{cases} 0, & s = 5, \\ -x_k, & s = 6, \\ -y_k, & s = 7. \end{cases}$$

The conservation laws, Eqs. (2.22), become

$$\sum_k \left( \frac{\partial L}{\partial \dot{x}_k} \xi_s^{\ k} + \frac{\partial L}{\partial \dot{y}_k} \eta_s^{\ k} + \frac{\partial L}{\partial \dot{z}_k} \zeta_s^{\ k} \right) = \text{constant} \qquad (s = 5, 6, 7)$$

or, using (2.40) and (2.35),

$$\sum_k (m_k \dot{x}_k y_k - m_k \dot{y}_k x_k) = \text{constant},$$

$$\sum_k (m_k \dot{x}_k z_k - m_k \dot{z}_k x_k) = \text{constant},$$

$$\sum_k (m_k \dot{y}_k z_k - m_k \dot{z}_k y_k) = \text{constant}.$$

In terms of the components of the momentum in the various coordinate directions, namely,

$$P_k^{(x)} = m_k \dot{x}_k, \qquad P_k^{(y)} = m_k \dot{y}_k, \qquad P_k^{(z)} = m_k \dot{z}_k \qquad (2.41)$$

(no summation on $k$), the conservation laws become

$$\sum_k (P_k^{(x)} y_k - P_k^{(y)} x_k) = \text{constant},$$

$$\sum_k (P_k^{(x)} z_k - P_k^{(z)} x_k) = \text{constant}, \qquad (2.42)$$

$$\sum_k (P_k^{(y)} z_k - P_k^{(z)} y_k) = \text{constant}.$$

These three formulas express the familiar facts that the $z$, $y$, and $x$ components of the total angular momentum are constants, respectively.

Invariance under the velocity transformations (2.33), which are sometimes called Galilean transformations, leads to a result which describes how the center of mass of the $n$-particle system moves. In addition, these velocity transformations provide us with our first example of transformations under which the fundamental integral is not absolutely invariant, but rather divergence-invariant or invariant up to a divergence term, in the sense of (2.14). In this case we make the same assumption as we made in the case of spatial translations, namely, that the potential energy depend only on the mutual positions of the particles; this assumption is expressed mathematically by (2.36). It is easily seen that such a potential is invariant under the transformations (2.33); more precisely we observe that

$$\begin{aligned}
V(\dots, &\bar{x}_i - \bar{x}_j, \bar{y}_i - \bar{y}_j, \bar{z}_i - \bar{z}_j, \dots) \\
&= V(\dots, (x_i + \varepsilon^8 t) - (x_j + \varepsilon^8 t), (y_i + \varepsilon^9 t) \\
&\quad - (y_j + \varepsilon^9 t), (z_i + \varepsilon^{10} t) - (z_j + \varepsilon^{10} t), \dots) \\
&= V(\dots, x_i - x_j, y_i - y_j, z_i - z_j, \dots). \qquad (2.43)
\end{aligned}$$

It is the invariance of the kinetic energy which gives rise to the divergence terms. We now compute a typical transformed velocity-squared term in the kinetic energy in order to obtain the exact form of these divergence terms.

$$\left(\frac{d\bar{x}_k}{d\bar{t}}\right)^2 = \left(\frac{d\bar{x}_k}{dt}\right)^2$$

$$= \left(\frac{d}{dt}(x_k + \varepsilon^8 t)\right)^2$$

$$= (\dot{x}_k + \varepsilon^8)^2$$

$$= \dot{x}_k^2 + 2\varepsilon^8 \dot{x}_k + (\varepsilon^8)^2$$

$$= \dot{x}_k^2 + \varepsilon^8 \frac{d}{dt}(2x_k) + o(\varepsilon^8).$$

Similarly

$$\left(\frac{d\bar{y}_k}{d\bar{t}}\right)^2 = \dot{y}_k^2 + \varepsilon^9 \frac{d}{dt}(2y_k) + o(\varepsilon^9)$$

and

$$\left(\frac{d\bar{z}_k}{d\bar{t}}\right)^2 = \dot{z}_k^2 + \varepsilon^{10} \frac{d}{dt}(2z_k) + o(\varepsilon^{10}).$$

Therefore,

$$\frac{1}{2}\sum_k m_k\left[\left(\frac{d\bar{x}_k}{d\bar{t}}\right)^2 + \left(\frac{d\bar{y}_k}{d\bar{t}}\right)^2 + \left(\frac{d\bar{z}_k}{d\bar{t}}\right)^2\right] - \frac{1}{2}\sum_k m_k(\dot{x}_k^2 + \dot{y}_k^2 + \dot{z}_k^2)$$

$$= \varepsilon^8 \frac{d}{dt}\sum_k m_k x_k + \varepsilon^9 \frac{d}{dt}\sum_k m_k y_k + \varepsilon^{10} \frac{d}{dt}\sum_k m_k z_k. \qquad (2.44)$$

We conclude from (2.43) and (2.44) that the difference between the Lagrangians in the "barred" and "unbarred" coordinate systems is

$$\bar{L} - L = \varepsilon^8 \frac{d}{dt}\Phi_8 + \varepsilon^9 \frac{d}{dt}\Phi_9 + \varepsilon^{10} \frac{d}{dt}\Phi_{10}, \qquad (2.45)$$

where

$$\Phi_8 = \sum_k m_k x_k, \qquad \Phi_9 = \sum_k m_k y_k, \qquad \Phi_{10} = \sum_k m_k z_k. \qquad (2.46)$$

Hence, (2.45) shows that the fundamental integral $\int (T - V)\, dt$ is divergence-invariant under (2.33) with divergence terms given by (2.46).

The group generators relative to (2.33) are

$$\tau_s \equiv \frac{\partial \bar{t}}{\partial \varepsilon^s}\bigg|_0 = 0 \qquad (s = 8, 9, 10),$$

$$\xi_s^{\ k} \equiv \frac{\partial \bar{x}_k}{\partial \varepsilon^s}\bigg|_0 = \begin{cases} t, & s = 8, \\ 0, & s = 9, 10; \end{cases} \qquad \eta_s^{\ k} \equiv \frac{\partial \bar{y}_k}{\partial \varepsilon^s}\bigg|_0 = \begin{cases} t, & s = 9, \\ 0, & s = 8, 10; \end{cases} \qquad (2.47)$$

$$\zeta_s^{\ k} \equiv \frac{\partial \bar{z}_k}{\partial \varepsilon^s}\bigg|_0 = \begin{cases} t, & s = 10, \\ 0, & s = 8, 9. \end{cases}$$

Whence, it follows from Eqs. (2.22) that

$$\sum_k \left( \frac{\partial L}{\partial \dot{x}_k} \xi_s^{\ k} + \frac{\partial L}{\partial \dot{y}_k} \eta_s^{\ k} + \frac{\partial L}{\partial \dot{z}_k} \zeta_s^{\ k} \right) - \Phi_s = \text{constant} \qquad (s = 8, 9, 10)$$

or, using (2.46), (2.47), and (2.35),

$$\sum_k m_k \dot{x}_k - \sum_k m_k x_k = \text{constant},$$

$$\sum_k m_k \dot{y}_k - \sum_k m_k y_k = \text{constant}, \qquad (2.48)$$

$$\sum_k m_k \dot{z}_k - \sum_k m_k z_k = \text{constant}.$$

To interpret the meaning of the first integrals in (2.48), let us denote

$$M = \sum_k m_k,$$

$$\qquad (2.49)$$

$$x_c = \frac{1}{M} \sum_k m_k x_k, \; y_c = \frac{1}{M} \sum_k m_k y_k, \; z_c = \frac{1}{M} \sum_k m_k z_k.$$

We remark that $(x_c, y_c, z_c)$ are the coordinates of the center of mass of the system, and $M$ is of course just the total mass of the $n$ particles. Next, from (2.28), we note that momentum is conserved, and as a result we obtain

$$\sum_k m_k \dot{x}_k = A_1, \qquad \sum_k m_k \dot{y}_k = A_2, \qquad \sum_k m_k \dot{z}_k = A_3 \qquad (2.50)$$

for constants $A_1, A_2$, and $A_3$. Denoting the constants on the right-hand sides of Eqs. (2.48) by $-B_1$, $-B_2$, and $-B_3$, respectively, we conclude from (2.48), (2.49), and (2.50) that

$$Mx_c = A_1 t + B_1,$$
$$My_c = A_2 t + B_2, \qquad (2.51)$$
$$Mz_c = A_3 t + B_3.$$

These equations tell how the center of mass moves; namely, it moves uniformly in time, or in different words, with constant velocity $\mathbf{v}_c$ given by

$$\mathbf{v}_c = \left(\frac{A_1}{M}, \frac{A_2}{M}, \frac{A_3}{M}\right).$$

All of the preceding results can be specialized to the fundamental integral

$$J = \int_{t_0}^{t_1} \left(\tfrac{1}{2} \sum_k m_k(\dot{x}_k^2 + \dot{y}_k^2 + \dot{z}_k^2) + \sum_{i<j} G\left(\frac{m_i m_j}{r_{ij}}\right)\right) dt, \qquad (2.52)$$

where $r_{ij}$ denotes the distance between the $i$th and $j$th particles and is defined in (2.39). Here the potential energy is

$$V = -\sum_{i<j} G\left(\frac{m_i m_j}{r_{ij}}\right), \qquad (2.53)$$

where $G$ is the gravitational constant. Therefore, (2.52) represents the variational integral for a system of $n$ particles where the forces between the particles are given by the inverse square law, or gravitational forces. Clearly, the potential energy defined by (2.53) satisfies the conditions mentioned earlier which guarantee the various conservation laws. To summarize our results, we state the following theorem.

**2.4  Theorem**  The fundamental integral (2.52) of the $n$-body system is divergence-invariant under the ten-parameter Galilean group given by (2.30)–(2.33); for this system, the total energy, linear momentum, and angular momentum are conserved, and the center of mass moves with constant velocity.  □

## 2.6  BIBLIOGRAPHIC NOTES

The original work on invariant variation problems was done by Noether [1] in 1918. This monumental paper can still be read for insight and motivation into these problems. Her paper has recently been translated into English by Tavel (see Noether [2]) who also supplies a brief motivation and historical sketch.

The approach taken in this monograph relies heavily on a set of fundamental invariance identities which, unlike the Noether identities, involves only first-order derivatives of the extremal curve; and they follow as a simple exercise in tensor analysis from the definition of invariance. No use is made of the Fundamental Variational Formula involving "small" variations of the dependent and independent variables. Our approach is due to Rund [4], although the fundamental invariance identities also appear in Trautman [1], but in the notation and formalism of modern differential geometry (bundles, jets, etc.). Also, this approach to the Noether theorem has recently appeared in Lovelock and Rund [1].

The classical variational method of obtaining the Noether theorem is described in several places. Among the most accessible are the books by Courant and Hilbert [1], Funk [1], Gelfand and Fomin [1], Rund [1], Sagan [1], and the paper by Hill [1], the latter of which is a common source quoted by physicists. A version of the Noether theorem using the language of modern differential geometry can be found in Komorowski [1] and García [1], as well as the article by Trautman [1] mentioned previously. A recent monograph by Marsden [1] also discusses a modern version of the theorem. The Noether theorem for discrete systems is given in an article by Logan [2], and the book by Edelen [1] discusses the theorem in the general setting of nonlocal variations.

The most important application of the Noether theorem for single integrals is to the problem of $n$-bodies. This application was first carried out by Bessel-Hagen [1] in 1921. A discussion of the $n$-body problem and interpretations of the first integrals and so on can be found in Whittaker [1].

There have been a great number of papers written in the last three decades on the Noether theorem. We have decided not to list a complete bibliography, but rather give only some popular or standard references.

## EXERCISES

**2-1**  Directly from the definition show that $J = \int_a^b t\dot{x}^2 \, dt$ is not invariant under the translation

$$\bar{t} = t + \varepsilon, \qquad \bar{x} = x.$$

Use the fundamental invariance identities to reach the same conclusion. However, show that $J$ is absolutely invariant under the translation

$$\bar{t} = t, \qquad \bar{x} = x + \varepsilon$$

and from Noether's theorem show that a first integral is given by

$$t\dot{x} = \text{constant}.$$

**2-2**  Consider the functional

$$J = \int_{t_0}^{t_1} \tfrac{1}{2}(m\dot{x}^2 - kx^2) \, dt$$

corresponding to a simple harmonic oscillator. Use Noether's theorem to derive the law of conservation of energy. Is momentum conserved?

**2-3** A particle of mass $m$ moves in one dimension under the influence of a force

$$F(x, t) = \frac{k}{x^2} e^{-t/\tau},$$

where $x$ is position, $t$ is time, and $k, \tau > 0$. Formulate Hamilton's principle for this system and derive the equation of motion; determine the Hamiltonian and compare it with the total energy. Is energy conserved?

**2-4** A particle of mass $m$ moves in a plane under the influence of a force $F = -Ar^{\alpha-1}, \alpha \neq 0, 1,$ and $A > 0$. Choose appropriate generalized coordinates and let the potential energy at the origin be zero. Find the Euler–Lagrange equations. Is energy conserved? Is the angular momentum about the origin conserved?

**2-5** Show that $(c - x^2)\sqrt{1 + \dot{x}^2} = 1, c$ a constant, is a first integral of the variational problem associated with the Lagrange function $L(t, x, \dot{x}) = x^2 + \sqrt{1 + \dot{x}^2}$.

**2-6** Show that the action integral for the simple pendulum is given by

$$J(\theta) = \int_{t_1}^{t_2} \left( \frac{m}{2} l^2 \dot{\theta}^2 + mgl \cos \theta \right) dt,$$

where $l$ is the length, $m$ is its mass, and $\theta$ is the angle of deflection. Derive the equation of motion. Is energy conserved? Is momentum conserved?

**2-7** Prove that if the fundamental integral $J = \int L(x, \dot{x}) \, dt$ is absolutely invariant under the "dilation" $\bar{t} = t + \gamma t$, then the Lagrangian must be homogeneous of degree 1 in the $\dot{x}$, i.e., $L(x, k\dot{x}) = kL(x, \dot{x}), k > 0$. Conclude that

$$L = \dot{x}^k \frac{\partial L}{\partial \dot{x}^k}.$$

**2-8** For the double pendulum shown below, show that the kinetic energy is given by

$$T = \tfrac{1}{2}(m_1 + m_2)a^2\dot{\phi}^2 + \tfrac{1}{2}m_2 b^2\dot{\psi}^2 + m_2 ab\dot{\phi}\dot{\psi} \cos(\phi - \psi)$$

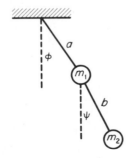

and the potential energy is given by

$$V = -(m_1 + m_2)ga \cos \phi - m_2 gb \cos \psi.$$

Formulate Hamilton's principle for this system and determine the equations of motion. What are the constants of the motion?

**2-9** (*Application to Optical Systems*)  According to Fermat's principle, light travels through a medium with index of refraction $\eta = \eta(x, y, z)$ in such a manner that

$$\delta \int_{t_2}^{t_1} \eta(x(t), y(t), z(t)) \sqrt{\dot{x}(t)^2 + \dot{y}(t)^2 + \dot{z}(t)^2}\, dt = 0.$$

(a)   Show that

$$\frac{d}{ds}\left[ \frac{\eta}{ds/dt} (\dot{x}, \dot{y}, \dot{z}) \right] = \operatorname{grad} \eta,$$

and consequently, the tangent to the ray path changes with arclength $s$ as a function of the gradient of the index (use the Euler–Lagrange equations).

(b)   Assuming that the system is invariant under translations parallel to the $x$-axis, apply the Noether theorem and thus derive Snell's law of reflection.

(c)   Assuming that the system is axially symmetric and thus invariant under rotations

$$\bar{x} = x \cos \theta - y \sin \theta,$$
$$\bar{y} = x \sin \theta + y \cos \theta,$$
$$\bar{z} = z,$$

show that an application of the Noether theorem leads to the Smith–Hemholtz invariant

$$\eta \beta x - \eta \alpha y = \text{constant},$$

where $\alpha$ and $\beta$ are the $x$ and $y$ direction cosines

$$\alpha \equiv \frac{x}{\sqrt{\dot{x}^2 + \dot{y}^2 + \dot{z}^2}}, \qquad \beta \equiv \frac{y}{\sqrt{\dot{x}^2 + \dot{y}^2 + \dot{z}^2}}$$

(see Blaker and Tavel [1]).

**2-10**   Prove that the fundamental invariance identities (2.15) are also sufficient for invariance.

# CHAPTER 3

# Generalized Killing Equations

## 3.1 INTRODUCTION

In the previous chapter we considered the group of transformations to be an a priori object. That is to say, given a group of invariance transformations and a Lagrangian, it is possible via the Noether theorem to directly write down explicit conservation laws for the system. Now we wish to consider the existence of the group of transformations as an a posteriori notion and show how a group can be determined under which a given variation problem is invariant. The importance of this latter problem is that once the group is determined, then conserved quantities can be written down directly by applying the Noether theorem.

To be more specific, let us examine the situation for single integrals. Let

$$J = \int_{t_0}^{t_1} L(t, x^1(t), \ldots, x^n(t), \dot{x}^1(t), \ldots, \dot{x}^n(t)) \, dt \qquad (3.1)$$

be a integral functional with the Lagrangian $L$ twice continuously differentiable in each of its $2n + 1$ arguments, and $x^k(t) \in C^2[t_0, t_1]$ for $k = 1, \ldots, n$.

According to Theorem 2.3, if the fundamental integral (3.1) is absolutely invariant under the $r$-parameter group of transformations

$$\bar{t} = t + \tau_s(t, x)\varepsilon^s, \qquad \bar{x}^k = x^k + \xi_s^{\ k}(t, x)\varepsilon^s, \qquad (3.2)$$

then the $r$ invariance identities

$$\frac{\partial L}{\partial t}\tau_s + \frac{\partial L}{\partial x^k}\xi_s^{\ k} + \frac{\partial L}{\partial \dot{x}^k}\left(\frac{d\xi_s^{\ k}}{dt} - \dot{x}^k\frac{d\tau_s}{dt}\right) + L\frac{d\tau_s}{dt} = 0 \qquad (3.3)$$

$(s = 1, \ldots, r)$ hold true. Upon expanding the exact derivatives in (3.3) it clearly follows that

$$\frac{\partial L}{\partial t}\tau_s + \frac{\partial L}{\partial x^k}\xi_s^{\ k} + \frac{\partial L}{\partial \dot{x}^k}\left(\frac{\partial \xi_s^{\ k}}{\partial t} + \frac{\partial \xi_s^{\ k}}{\partial x^j}\dot{x}^j - \dot{x}^k\frac{\partial \tau_s}{\partial t} - \dot{x}^k\dot{x}^j\frac{\partial \tau_s}{\partial x^j}\right)$$

$$+ L\left(\frac{\partial \tau_s}{\partial t} + \frac{\partial \tau_s}{\partial x^j}\dot{x}^j\right) = 0. \qquad (3.4)$$

Heretofore, these identities have been regarded as a condition involving the Lagrangian $L$ and the quantities $\tau_s$ and $\xi_s^{\ k}$ that are specified uniquely by the given infinitesimal group of transformations (3.2). We shall now alter our viewpoint. As we remarked in Chapter 2, Eqs. (3.4) are identities in $(t, x^k)$ for arbitrary directional arguments $\dot{x}^k$; implicit in this remark is the fact that the group generators $\tau_s$ and $\xi_s^{\ k}$ depend only on $t$ and $x^k$ and not on the $\dot{x}^k$. Therefore, we can regard the identities (3.4) as a set of partial differential equations in unknowns $\tau_s$ and $\xi_s^{\ k}$. And, due to the arbitrariness of the $\dot{x}^k$, we can further reduce (3.4) to obtain a system of first-order partial differential equations in $\tau_s$ and $\xi_s^{\ k}$ by equating to zero the coefficients of the powers of the $\dot{x}^k$. It is obvious that if this reduced system admits a solution, then first integrals may be obtained immediately using the Noether theorem; for, we have determined a group under which the fundamental integral is invariant.

### 3.2  EXAMPLE—THE EMDEN EQUATION

In this section we illustrate the general method outlined in the preceding introduction by calculating a first integral of the classical Emden equation

$$\frac{d^2x}{dt^2} + \frac{2}{t}\frac{dx}{dt} + x^5 = 0. \qquad (3.5)$$

This equation, first studied by the German astrophysicist Robert Emden in 1907, arises in the investigation of the thermal behavior of a spherical gas cloud acting under the mutual attraction of its molecules and subject to the laws of thermodynamics (see H. T. Davis [1], p. 371 ff). The variation integral which has the Emden equation as its associated Euler–Lagrange equation is

$$J = \int_{t_0}^{t_1} t^2 \left( \frac{\dot{x}^2}{2} - \frac{x^6}{6} \right) dt. \tag{3.6}$$

so that the Lagrangian $L(t, x, \dot{x})$ is given by

$$L = t^2 \left( \frac{\dot{x}^2}{2} - \frac{x^6}{6} \right). \tag{3.7}$$

We look for a one-parameter family of transformations

$$\bar{t} = t + \tau(t, x)\varepsilon, \qquad \bar{x} = x + \xi(t, x)\varepsilon \tag{3.8}$$

under which (3.6) is invariant. First, we note that

$$\frac{\partial L}{\partial t} = t(\dot{x}^2 - \tfrac{1}{3}x^6), \qquad \frac{\partial L}{\partial x} = -t^2 x^5, \qquad \frac{\partial L}{\partial \dot{x}} = t^2 \dot{x}.$$

Substituting these quantities into the invariance identity (3.4) (with $s = 1$, $k = 1$), we obtain

$$\left( \dot{x}^2 - \frac{x^6}{3} \right) \tau - tx^5 + t\dot{x} \left( \frac{\partial \xi}{\partial t} + \frac{\partial \xi}{\partial x} \dot{x} - \dot{x} \frac{\partial \tau}{\partial t} - \dot{x}^2 \frac{\partial \tau}{\partial x} \right)$$

$$+ t \left( \frac{\dot{x}^2}{2} - \frac{x^6}{6} \right) \left( \frac{\partial \tau}{\partial t} + \frac{\partial \tau}{\partial x} \dot{x} \right) = 0.$$

Collecting the coefficients of the powers of $\dot{x}$, namely, $\dot{x}^0$, $\dot{x}^1$, $\dot{x}^2$, and $\dot{x}^3$, and setting those coefficients equal to zero, we obtain the following system of partial differential equations for $\tau$ and $\xi$:

(i) $\dfrac{x}{3}\tau + t\xi + \dfrac{tx}{6}\dfrac{\partial \tau}{\partial t} = 0;$

(ii) $\dfrac{\partial \xi}{\partial t} - \dfrac{x^6}{6}\dfrac{\partial \tau}{\partial x} = 0;$

$\qquad\qquad\qquad\qquad\qquad\qquad\qquad\qquad\qquad\qquad (3.9)$

(iii) $t\tau + t\dfrac{\partial \xi}{\partial x} - \dfrac{t^2}{2}\dfrac{\partial \tau}{\partial t} = 0;$

(iv) $\dfrac{\partial \tau}{\partial x} = 0.$

From (3.9iv) it follows that $\tau = \tau(t)$ and, as a result, (3.9ii) shows that $\partial \xi / \partial t = 0$ or $\xi = \xi(x)$. Therefore, Eqs. (3.9) reduce to a pair of ordinary differential equations

$$\frac{x}{3}\tau + t\xi + \frac{tx}{6}\frac{d\tau}{dt} = 0,$$

$$\tau + t\frac{d\xi}{dx} - \frac{t}{2}\frac{d\tau}{dt} = 0,$$

which has, by simple inspection, solutions given by

$$\tau = t, \qquad \xi = -\tfrac{1}{2}x.$$

Consequently, as can easily be verified, $J$ is invariant under the transformation group

$$\bar{t} = t + t\varepsilon,$$
$$\bar{x} = x - \tfrac{1}{2}x\varepsilon.$$

From Theorem 2.3, the corollary of Noether's theorem, we obtain the first integral

$$\frac{t^3 x^6}{6} + \frac{t^3 \dot{x}^2}{2} + \frac{t^2 x \dot{x}}{2} = C$$

of Emden's equation.

## 3.3 KILLING'S EQUATIONS

We consider the fundamental integral

$$J(x^1, \ldots, x^n) = \int_{t_0}^{t_1} \tfrac{1}{2} g_{kl} \dot{x}^k \dot{x}^l \, dt \tag{3.10}$$

of a dynamical system having no external forces present, in which the $x^1, \ldots, x^n$ represent the generalized coordinates and $g_{kl} = g_{kl}(x^1, \ldots, x^n)$ are class $C^2$ functions which are symmetric in the indices $k$ and $l$. Here, the Lagrangian $L$ given by†

$$L = \tfrac{1}{2} g_{kl} \dot{x}^k \dot{x}^l \tag{3.11}$$

coincides with the kinetic energy of the system. We seek to determine a one-parameter transformation of the form

$$\bar{t} = t, \qquad \bar{x}^k = x^k + \xi^k(x)\varepsilon \tag{3.12}$$

† For simplicity, assume $g_{kl}$ are constant.

under which the fundamental integral (3.10) is invariant. To this end, we shall write down the invariance identities (3.4) for the present system. First, we note from (3.11) that

$$\frac{\partial L}{\partial t} = 0, \qquad \frac{\partial L}{\partial x^j} = \frac{1}{2} \frac{\partial g_{kl}}{\partial x^j} \dot{x}^k \dot{x}^l = 0$$

and

$$\frac{\partial L}{\partial \dot{x}^j} = \frac{1}{2} g_{kl}(\dot{x}^k \delta_j^{\ l} + \dot{x}^l \delta_j^{\ k}).$$

Substitution of these quantities into (3.4) yields

$$\frac{1}{2} g_{kl}(\dot{x}^k \delta_j^{\ l} + \dot{x}^l \delta_j^{\ k}) \left( \frac{\partial \xi^j}{\partial x^i} \dot{x}^i \right) = 0,$$

where we have also used the fact that $\tau \equiv 0$ and $\partial \xi / \partial t = 0$. Upon simplification, this becomes

$$\frac{1}{2} g_{kl} \dot{x}^k \dot{x}^i \frac{\partial \xi^l}{\partial x^i} + \frac{1}{2} g_{kl} \dot{x}^l \dot{x}^i \frac{\partial \xi^k}{\partial x^i} = 0.$$

By renaming indices, we obtain

$$\frac{1}{2} \left( g_{ik} \frac{\partial \xi^k}{\partial x^l} + g_{kl} \frac{\partial \xi^k}{\partial x^i} \right) \dot{x}^i \dot{x}^l = 0.$$

Since the $\dot{x}^i \dot{x}^l$ are arbitrary, it follows that

$$g_{ik} \frac{\partial \xi^k}{\partial x^l} + g_{kl} \frac{\partial \xi^k}{\partial x^i} = 0 \qquad (3.13)$$

for all $i, l = 1, \ldots, n$. Therefore, if the functional (3.10) is to be invariant under (3.12), then the generators $\xi^k$ in (3.12) must satisfy the system of first-order partial differential equations (3.13). If Eqs. (3.13) can be solved to obtain the $\xi^k$, then Noether's theorem guarantees the first integral

$$\frac{\partial L}{\partial \dot{x}^k} \xi^k = C$$

of the governing equations of motion which are, in this case,

$$g_{kj} \ddot{x}^j = 0.$$

Equations (3.13), known as Killing's equations, were first obtained by W. Killing [1] in 1892 in the context of describing the motions of an $n$-dimensional manifold with fundamental form given by (3.11) (see Eisenhart [1]). In general, we shall refer to as *Killing's equations* the system of linear homogeneous partial differential equations in $\tau_s$ and $\xi_s{}^k$ obtained from the invariance identities (3.4) by equating to zero the coefficients of the powers of the directional arguments $\dot{x}^k$. Implicit in this remark is the assumption that the identities (3.4) can be written, after substitution of $\partial L/\partial t$, $\partial L/\partial x^k$, and $\partial L/\partial \dot{x}^k$, as a polynomial in the $\dot{x}^k$. If these general Killing equations can be solved for $\tau_s$ and $\xi_s{}^k$, then Noether's theorem gives explicit first integrals

$$\left( L - \dot{x}^k \frac{\partial L}{\partial \dot{x}^k} \right) \tau_s + \frac{\partial L}{\partial \dot{x}^k} \xi_s{}^k = C$$

of the Euler–Lagrange equations for the system.

One should not conclude from the preceding remarks that the technique described represents a universal method for obtaining first integrals of a given system of second-order ordinary differential equations. It may be the case that the general system of Killing equations is not easily solvable, or there may exist no nontrivial solutions.

**3.1   Example**   The differential equation

$$\ddot{x} + (t + 1)x - 1 = 0$$

is the Euler–Lagrange equation corresponding to the fundamental integral

$$J = \int_{t_0}^{t_1} (\dot{x}^2 + (t + 1)x^2 + 2x) \, dt.$$

Using the method described above and applied to the Emden equation in Section 3.2, we arrive at the set of Killing equations

$$x\tau + (tx + 4)\xi + (x^2 + tx^2 + 2x) \frac{d\tau}{dt} = 0,$$

$$2 \frac{d\xi}{dx} - \frac{d\tau}{dt} = 0,$$

where $\tau = \tau(t)$ and $\xi = \xi(x)$. It is not difficult to see that this system admits only the trivial solution $\xi = \tau = 0$.

## 3.4   THE DAMPED HARMONIC OSCILLATOR

In this section we shall show how the method described in the preceding sections can be applied to a dissipative system. We consider (see Fig. 5) a harmonic oscillator with restoring force $-kx$ which is emersed in a liquid in such a way that the motion of the mass $m$ is damped by a force proportional

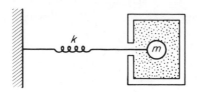

**FIGURE 5**

to its velocity. Upon writing Newton's second law for the system, we obtain the governing equation of motion

$$m\ddot{x} + a\dot{x} + kx = 0, \tag{3.14}$$

where $k > 0$, $a > 0$. It is not difficult to see that the Lagrangian which leads to (3.14) is

$$L(t, x, \dot{x}) = \tfrac{1}{2}(m\dot{x}^2 - kx^2)e^{(a/m)t}. \tag{3.15}$$

In order to determine a first integral of (3.14) we shall try to find a one-parameter family of transformations of the form

$$\bar{t} = t + \varepsilon\tau(t, x), \qquad \bar{x} = x + \varepsilon\xi(t, x) \tag{3.16}$$

under which the action integral is invariant. From our previous remarks the invariance identity

$$L_t\tau + L_x\xi + L_{\dot{x}}(\dot{\xi} - \dot{x}\dot{\tau}) + L\dot{\tau} = 0 \tag{3.17}$$

must hold true. For the present case we have

$$L_t = \tfrac{1}{2}(m\dot{x}^2 - kx^2)\frac{a}{m}\,e^{(a/m)t}, \qquad L_x = -kxe^{(a/m)t}, \qquad L_{\dot{x}} = m\dot{x}e^{(a/m)t}.$$

Upon substituting these quantities into (3.17) and expanding the total derivatives $\dot{\xi}$ and $\dot{\tau}$, we obtain

$$\frac{a}{2m}(m\dot{x}^2 - kx^2)\tau - kx\xi + m\dot{x}\left(\frac{\partial\xi}{\partial t} + \frac{\partial\xi}{\partial x}\dot{x} - \dot{x}\frac{\partial\tau}{\partial t} - \dot{x}^2\frac{\partial\tau}{\partial x}\right)$$

$$+ \tfrac{1}{2}(m\dot{x}^2 - kx^2)\left(\frac{\partial\tau}{\partial t} + \frac{\partial\tau}{\partial x}\dot{x}\right) = 0.$$

When the coefficients of $\dot{x}^0$, $\dot{x}$, $\dot{x}^2$, and $\dot{x}^3$ are collected and equated to zero, we obtain the following system of four first-order partial differential equations, the generalized Killing equations:

(i)   $\dfrac{a}{2m}\,\tau x + \xi + \dfrac{x}{2}\dfrac{\partial\tau}{\partial t} = 0;$

(ii)   $m\dfrac{\partial\xi}{\partial t} - \dfrac{k}{2}\dfrac{\partial\tau}{\partial x}x^2 = 0;$

(iii)   $\dfrac{a}{2m}\,\tau + \dfrac{\partial\xi}{\partial x} - \dfrac{1}{2}\dfrac{\partial\tau}{\partial t} = 0;$

(iv)   $\dfrac{\partial\tau}{\partial x} = 0.$

$$(3.18)$$

From (3.18iv) we can conclude that $\tau = \tau(t)$ and consequently from (3.18ii) it follows that $\xi = \xi(x)$. Therefore these four equations reduce to two ordinary differential equations

(i)   $\dfrac{a}{2m}\,\tau x + \xi + \dfrac{x}{2}\dfrac{d\tau}{dt} = 0,$

(ii)   $\dfrac{a}{2m}\,\tau + \dfrac{d\xi}{dx} - \dfrac{1}{2}\dfrac{d\tau}{dt} = 0.$

$$(3.19)$$

If the second equation in (3.19) is multiplied by $x$ and added to the first equation, then we obtain

$$\frac{a}{m}\,\tau x + \xi + x\frac{d\xi}{dx} = 0,$$

which implies that $\tau = \text{constant} = c$. In that case we observe from (3.19ii) that $\xi = -(ac/2m)x$. Upon choosing $c = 1$ we therefore obtain the one-parameter family of transformations

$$\bar{t} = t + \varepsilon, \qquad \bar{x} = x - \varepsilon\frac{ax}{2m}, \tag{3.20}$$

under which, by their very construction, the action integral is invariant.

Hence by Noether's theorem there is a first integral or constant of the motion; i.e.,

$$(L - \dot{x}L_{\dot{x}})\tau + L_{\dot{x}}\xi = \text{constant},$$

or, in the present case,

$$(m\dot{x}^2 + kx^2 + ax\dot{x})e^{(a/m)t} = \text{constant}.$$

## 3.5   THE INVERSE PROBLEM

It is clear from the preceding discussion that in order to apply group theoretic methods to calculate first integrals of given differential equations, it is necessary to be able to compute the Lagrange function $L$ from the given differential equation. That is, given a second-order differential equation

$$\ddot{x} = F(t, x, \dot{x}), \tag{3.21}$$

we might ask which Lagrangian $L$ leads to (3.21) as a necessary condition for $\int L \, dt \to \text{Ext}$; in other words, for which function $L = L(t, x, \dot{x})$ does (3.21) coincide with the Euler–Lagrange equation

$$\frac{\partial L}{\partial x} - \frac{d}{dt}\frac{\partial L}{\partial \dot{x}} = 0. \tag{3.22}$$

This is the so-called inverse problem of the calculus of variations, and we shall now indicate its solution for the simplest variational problem.

If we rewrite (3.22) by expanding the total derivative, we obtain (using the subscript notation for partial differentiation)

$$L_x - L_{\dot{x}t} - L_{\dot{x}x}\dot{x} - L_{\dot{x}\dot{x}}\ddot{x} = 0$$

or, using (3.21),

$$L_x - L_{\dot{x}t} - L_{\dot{x}x}\dot{x} - L_{\dot{x}\dot{x}}F = 0,$$

the latter equation being an identity in $t$, $x$, and $\dot{x}$. If this equation is differentiated with respect to $\dot{x}$, it becomes

$$L_{\dot{x}t\dot{x}} + \dot{x}L_{\dot{x}x\dot{x}} + L_{\dot{x}\dot{x}}F_{\dot{x}} + L_{\dot{x}\dot{x}\dot{x}}F = 0. \tag{3.23}$$

Here, of course, we are assuming that the Lagrangian $L$ has continuous partial derivatives of the third order. From (3.23) we can find a differential equation for $u = u(t, x, \dot{x})$, where

$$u = L_{\dot{x}\dot{x}}. \tag{3.24}$$

The equation is

$$u_t + \dot{x}u_x + Fu_{\dot{x}} + F_{\dot{x}}u = 0. \tag{3.25}$$

Equation (3.25) is a first-order linear partial differential equation in the unknown $u$. It can be solved, for example, by the method of characteristics (see Courant and Hilbert [2]). Once the solution $u$ is known, then the Lagrangian $L$ can be determined from (3.24) by simple quadrature.

## EXERCISES

**3-1**  Determine all functionals $J = \int_a^b L(t, x, \dot{x})\, dt$ for which the extremals are the straight lines $x = At + B$.

**3-2**  Consider the nonlinear ordinary differential equation

$$\ddot{x} + x^3 = 0.$$

(a)  Find a Lagrangian $L = L(t, x, \dot{x})$ for which this differential equation coincides with the Euler–Lagrange equation for the functional

$$J(x) = \int L(t, x, \dot{x})\, dt.$$

(b)  Find all transformations of the form

$$\bar{t} = t + \varepsilon\tau(t, x), \qquad \bar{x} = x + \varepsilon\xi(t, x)$$

under which $J(x)$ is absolutely invariant.

(c)  Apply Noether's theorem to find a first integral of the given differential equation.

**3-3**  Let $F: R^n \to R^1$ be a real-valued $C^\infty$ function on $R^n$, and let

$$\bar{x}^k = \phi^k(x^1, \ldots, x^n; \varepsilon^1, \ldots, \varepsilon^r) \qquad (k = 1, \ldots, n)$$

be an $r$-parameter family of bijective, continuously differentiable transformations on $R^n$ such that to the values $\varepsilon^s = 0$ (for all $s$) of the parameter corresponds the identity transformation. We say $F$ is invariant under the transformations $\phi^k$ if and only if

$$\left. \frac{\partial}{\partial \varepsilon^s} (F \circ \Phi) \right|_{\varepsilon = 0} = 0,$$

where $\Phi = (\phi^1, \ldots, \phi^n)$.

(a)  Prove that if the function $F$ is invariant under the transformations $\phi^k$, then $F$ must satisfy the first-order system of $r$ linear partial differential equation

$$\frac{\partial F}{\partial x^1}\, \tau_s{}^1(x) + \cdots + \frac{\partial F}{\partial x^n}\, \tau_s{}^n(x) = 0 \qquad (s = 1, \ldots, r),$$

where

$$\tau_s{}^k(x) = \left. \frac{\partial \phi^k}{\partial \varepsilon^s} \right|_{\varepsilon = 0}.$$

(b)   If $F: R^2 \to R^1$ is invariant under the rotation

$$\bar{x}^1 = x^1 + \varepsilon x^2 + o(\varepsilon), \qquad \bar{x}^2 = x^2 - \varepsilon x^1 + o(\varepsilon),$$

prove that $F$ must be constant on the circles $(x^1)^2 + (x^2)^2 = c$.

(c)   Prove that the only function $F: R^2 \to R^1$ which is invariant under the two-parameter family of translations

$$\bar{x}^1 = x^1 + \varepsilon^1, \qquad \bar{x}^2 = x^2 + \varepsilon^2$$

is the constant function.

(d)   Prove that if $F: R^n \to R^1$ is invariant under the $r$-parameter family of transformations $\Phi$, and if $F$ is not constant, then $r < n$.

(e)   Prove:   If $F: R^n \to R^1$ satisfies the system

$$\sum_{k=1}^{n} \frac{\partial F}{\partial x^k} \tau_s^k(x) = 0 \qquad (s = 1, \ldots, r),$$

where the $\tau_s^k(s)$ are given, then there exists an $r$-parameter family of transformations on $R^n$ under which $F$ is invariant.

**3-4**   Find a functional whose Euler–Lagrange equation is

$$\ddot{x} = 6t.$$

# CHAPTER 4

# Invariance of Multiple Integrals

## 4.1 BASIC DEFINITIONS

The concepts in Sections 2.1–2.4 have an obvious analogy in the multiple integral case. In Section 1.5 we stated a necessary condition for a multiple integral to be extremal. We shall now seek necessary conditions for a multiple integral to be invariant under a given $r$-parameter family of transformations.

As in Section 1.5, we consider a fundamental multiple integral of the form

$$J(x) = \int_D L\left(t, x(t), \frac{\partial x(t)}{\partial t}\right) dt^1 \cdots dt^m, \tag{4.1}$$

where $D \subseteq R^m$, $t = (t^1, \ldots, t^m)$, $x(t) = (x^1(t), \ldots, x^n(t))$, and $\partial x(t)/\partial t$ denotes the collection of first partial derivatives $\dot{x}_\alpha{}^k = \partial x^k/\partial t^\alpha$. Here, and in the sequel, lower case Greek indices $\alpha, \beta, \ldots$ will have the range $1, \ldots, m$ and the lower case Latin indices $i, j, k, \ldots$ will have the range $1, \ldots, n$. The differentiability conditions on $L$ and the functions $x^k(t)$ are the same as assumed in Section 1.5.

We assume that there is given an $r$-parameter family of transformations on the variables $t^1, \ldots, t^m, x^1, \ldots, x^n$ of the form

$$\bar{t}^\alpha = \phi^\alpha(t, x, \varepsilon), \qquad \bar{x}^k = \psi^k(t, x, \varepsilon), \tag{4.2}$$

where $\varepsilon = (\varepsilon^1, \ldots, \varepsilon^r)$ denotes the set of independent parameters which belong to some open rectangle in $R^r$ containing the origin. We assume that the functions $\phi^\alpha$ and $\psi^k$ in (3.2) are of class $C^2$ in all their arguments and that the zero value of all the parameters, i.e., $\varepsilon^1 = \cdots = \varepsilon^r = 0$, gives the identity transformation

$$\bar{t}^\alpha = \phi^\alpha(t, x, 0) = t^\alpha, \qquad \bar{x}^k = \psi^k(t, x, 0) = x^k.$$

By Taylor's theorem, the right-hand sides of (4.2) can be expanded about $\varepsilon = 0$ to obtain

$$\bar{t}^\alpha = t^\alpha + \tau_s{}^\alpha(t, x)\varepsilon^s + o(\varepsilon), \qquad \bar{x}^k = x^k + \xi_s{}^k(t, x)\varepsilon^s + o(\varepsilon). \qquad (4.3)$$

Here, and in the subsequent discussion, the index $s$ will range over $1, \ldots, r$. The *generators* $\tau_s{}^\alpha(t, x)$ and $\xi_s{}^k(t, x)$ of the transformation are given by

$$\tau_s{}^\alpha(t, x) = \frac{\partial \phi^\alpha}{\partial \varepsilon^s}(t, x, 0), \qquad \xi_s{}^k(t, x) = \frac{\partial \psi^k}{\partial \varepsilon^s}(t, x, 0). \qquad (4.4)$$

For $\varepsilon$ sufficiently close to the origin in $R^r$, it can be shown (see Exercise 4.2) that the transformation (4.2) carries a surface $C_m$ with equation

$$x^k = x^k(t), \quad t \in D \qquad (k = 1, \ldots, n)$$

into an $r$-parameter family of surfaces $\bar{C}_m$ in $(\bar{t}^1, \ldots, \bar{t}^m, \bar{x}^1, \ldots, \bar{x}^n)$ space with equation

$$\bar{x}^k = \bar{x}^k(\bar{t}), \quad \bar{t} \in \bar{D} \qquad (k = 1, \ldots, n).$$

This construction is similar to the one presented in Section 2.1: we subject $C_m$ to the transformation (4.2) to obtain

$$\bar{t}^\alpha = \phi^\alpha(t, x(t), \varepsilon), \qquad \bar{x}^k = \psi^k(t, x(t), \varepsilon).$$

For sufficiently small $\varepsilon$ the first equation may be solved for $t^\alpha$ to obtain

$$t^\alpha = T^\alpha(\bar{t}, \varepsilon).$$

Upon substitution of these quantities into the second equation we obtain

$$\bar{x}^k = \psi^k(T(\bar{t}, \varepsilon), x(T(\bar{t}, \varepsilon)), \varepsilon)$$
$$\equiv \bar{x}^k(\bar{t}).$$

Therefore, the functions $x^k = x^k(t)$ and the functions $\bar{x}^k = \bar{x}^k(\bar{t})$ are related by means of the transformation (4.2) via the condition

$$\psi^k(t, x(t), \varepsilon) = \bar{x}^k(\phi(t, x(t), \varepsilon)), \qquad (4.5)$$

where $\phi = (\phi^1, \ldots, \phi^m)$. Clearly, $\bar{x}(\bar{t})$ is defined on $\bar{D}$, where

$$\bar{D} = \{\phi(t, x(t), \varepsilon) : t \in D\}.$$

As one additional observation, we note that in order for $C_m$ to be an $m$-dimensional hypersurface in $R^{m+n}$ it is sufficient that

$$\text{rank}\left(\frac{\partial x^k}{\partial t^\alpha}\right) = m.$$

We shall adhere to this assumption in the subsequent discussion.

We can now define what is inferred by stating that the multiple integral (4.1) is invariant under the $r$-parameter family of transformations given by (4.2).

**4.1  Definition**  The fundamental integral (4.1) is absolutely invariant under the $r$-parameter family of transformations (4.2) if and only if given any hypersurface $x = x(t)$ of class $C_n{}^2(D)$ and any rectangle $G \subseteq D$, we have

$$\int_{\bar{G}} L\left(\bar{t}, \bar{x}(\bar{t}), \frac{\partial \bar{x}(\bar{t})}{\partial \bar{t}}\right) d\bar{t} - \int_G L\left(t, x(t), \frac{\partial x(t)}{\partial t}\right) dt = o(\varepsilon) \tag{4.6}$$

for every $\varepsilon$ sufficiently close to the origin in $R^r$, and where

$$\bar{G} = \{\bar{t} \in R^m | \bar{t} = \phi(t, x(t), \varepsilon), t \in G\}.$$

**4.1  Remark**  Condition (4.6) can be formulated completely in terms of the Lagrange function $L$ without reference to the fundamental integral. In fact, (4.6) may be replaced by the equation

$$L\left(\bar{t}, \bar{x}(\bar{t}), \frac{\partial \bar{x}(\bar{t})}{\partial \bar{t}}\right) \det\left(\frac{\partial \bar{t}^\alpha}{\partial t^\beta}\right) - L\left(t, x(t), \frac{\partial x(t)}{\partial t}\right) = o(\varepsilon). \tag{4.7}$$

To verify the correctness of (4.7) we note that a change of variables in the first integral in (4.6) yields

$$\int_{\bar{G}} L\left(\bar{t}, \bar{x}(\bar{t}), \frac{\partial \bar{x}(\bar{t})}{\partial \bar{t}}\right) d\bar{t} = \int_G L\left(\bar{t}, \bar{x}(\bar{t}), \frac{\partial \bar{x}(\bar{t})}{\partial \bar{t}}\right) \det\left(\frac{\partial \bar{t}^\alpha}{\partial t^\beta}\right) dt,$$

where by $\bar{t}$, in the integral on right, we mean $\bar{t} = \phi(t, x(t), \varepsilon)$. The arbitrariness of the region $G$ then easily yields (4.7).

**4.1  Example**  Before proceeding with a derivation of invariance identities and conservation laws in the multiple integral case we illustrate the preceding definitions with a specific example. We consider the functional

$$J(u(t, x)) = \int_{t_0}^{t_1} \int_{x_0}^{x_1} \tfrac{1}{2}(u_t{}^2 - c^2 u_x{}^2) \, dt \, dx, \tag{4.8}$$

where $u_t = \partial u / \partial t$ and $u_x = \partial u / \partial x$, and $c$ is a constant. The Lagrangian is given by

$$L(t, x, u, u_t, u_x) = \tfrac{1}{2}(u_t{}^2 - c^2 u_x{}^2), \tag{4.9}$$

and according to (1.33), the Euler–Lagrange equation is the wave equation

$$c^2 u_{xx} - u_{tt} = 0.$$

We shall now give a direct proof that the functional defined by (4.8) is absolutely invariant under the four-parameter family of transformations

$$\bar{t} = t + \varepsilon^1 + \varepsilon^3 x, \qquad \bar{x} = x + \varepsilon^2 + \varepsilon^3 c^2 t, \qquad \bar{u} = u + \varepsilon^4, \qquad (4.10)$$

where $\varepsilon^1, \ldots, \varepsilon^4$ are the parameters of the group. The reader should alert himself that a superscript can denote either an exponent or an index, depending on the context.

To show absolute invariance of (4.8) we must show, according to Definition 4.1,

$$L(\bar{t}, \bar{x}, \bar{u}, \bar{u}_{\bar{t}}, \bar{u}_{\bar{x}})\Delta - L(t, x, u, u_t, u_x) = o(\varepsilon), \qquad (4.11)$$

where $\Delta$ denotes the Jacobian of the transformation $(x, t) \rightarrow (\bar{x}, \bar{t})$. Clearly, from (4.10),

$$\Delta = \det \begin{bmatrix} \dfrac{\partial \bar{t}}{\partial t} & \dfrac{\partial \bar{x}}{\partial t} \\[2ex] \dfrac{\partial \bar{t}}{\partial x} & \dfrac{\partial \bar{x}}{\partial x} \end{bmatrix} = 1 - c^2(\varepsilon^3)^2. \qquad (4.12)$$

It remains to calculate the derivatives $\bar{u}_{\bar{t}}$ and $\bar{u}_{\bar{x}}$. Fortunately, it is not necessary to calculate the function $\bar{u} = \bar{u}(\bar{t}, \bar{x})$; the derivatives can be computed directly from the chain rule much the same as was done in Section 2.2. To this end, we have

$$\begin{bmatrix} \dfrac{\partial u}{\partial t} \\[2ex] \dfrac{\partial u}{\partial x} \end{bmatrix} = \begin{bmatrix} \dfrac{\partial \bar{u}}{\partial t} \\[2ex] \dfrac{\partial \bar{u}}{\partial x} \end{bmatrix} = \begin{bmatrix} \dfrac{\partial \bar{t}}{\partial t} & \dfrac{\partial \bar{x}}{\partial t} \\[2ex] \dfrac{\partial \bar{t}}{\partial x} & \dfrac{\partial \bar{x}}{\partial x} \end{bmatrix} \begin{bmatrix} \dfrac{\partial \bar{u}}{\partial \bar{t}} \\[2ex] \dfrac{\partial \bar{u}}{\partial \bar{x}} \end{bmatrix}$$

or

$$\begin{bmatrix} \dfrac{\partial u}{\partial t} \\[2ex] \dfrac{\partial u}{\partial x} \end{bmatrix} = \begin{bmatrix} 1 & c^2 \varepsilon^3 \\[1ex] \varepsilon^3 & 1 \end{bmatrix} \begin{bmatrix} \dfrac{\partial \bar{u}}{\partial \bar{t}} \\[2ex] \dfrac{\partial \bar{u}}{\partial \bar{x}} \end{bmatrix}.$$

Upon inverting the Jacobi matrix on the right-hand side, we obtain

$$
\begin{bmatrix} \dfrac{\partial \bar{u}}{\partial \bar{t}} \\[2ex] \dfrac{\partial \bar{u}}{\partial \bar{x}} \end{bmatrix} = \frac{1}{\Delta} \begin{bmatrix} 1 & -c^2 \varepsilon^3 \\[1ex] -\varepsilon^3 & 1 \end{bmatrix} \begin{bmatrix} \dfrac{\partial u}{\partial t} \\[2ex] \dfrac{\partial u}{\partial x} \end{bmatrix}
$$

or, in component form,

$$
\bar{u}_{\bar{t}} = \frac{1}{\Delta}(u_t - c^2 \varepsilon^3 u_x),
$$

$$
\bar{u}_{\bar{x}} = \frac{1}{\Delta}(u_x - \varepsilon^3 u_t).
$$

(4.13)

Substituting (4.13) and (4.12) into (4.11), where $L$ is given by (4.9), we obtain, after some simplification,

$$
\tfrac{1}{2}(\bar{u}_{\bar{t}}^{\,2} - c^2 \bar{u}_{\bar{x}}^{\,2})\Delta - \tfrac{1}{2}(u_t^{\,2} - c^2 u_x^{\,2})
$$

$$
= \frac{1}{2\Delta}(u_t^{\,2} - c^2 u_x^{\,2} + c^4(\varepsilon^3)^2 u_x^{\,2} - c^2(\varepsilon^3)^2 u_t^{\,2}) - \tfrac{1}{2}(u_t^{\,2} - c^2 u_x^{\,2})
$$

$$
= \frac{1}{2\Delta}(u_t^{\,2} - c^2 u_x^{\,2})\Delta - \tfrac{1}{2}(u_t^{\,2} - c^2 u_x^{\,2})
$$

$$
= 0.
$$

Therefore, according to Definition 2.1, the functional defined by (4.8) is absolutely invariant under the four-parameter group of transformations (4.10).

## 4.2  THE FUNDAMENTAL THEOREMS

Now we shall show that the invariance of the fundamental integral under an $r$-parameter group of transformations implies the existence of a set of $r$ differential identities.

**4.1  Theorem**  A necessary condition for the fundamental integral (4.1) to be absolutely invariant under the $r$-parameter family of transformations (4.2) is that the Lagrangian $L(t, x, \partial x/\partial t)$ and its derivatives satisfy the $r$ identities

$$
\frac{\partial L}{\partial t^\alpha}\tau_s^{\,\alpha} + \frac{\partial L}{\partial x^k}\xi_s^{\,k} + \frac{\partial L}{\partial \dot{x}_\alpha^{\,k}}\left(\frac{d\xi_s^{\,k}}{dt^\alpha} - \dot{x}_\beta^{\,k}\frac{\partial \tau_s^{\,\beta}}{dt^\alpha}\right) + L\frac{d\tau_s^{\,\alpha}}{dt^\alpha} = 0, \quad (4.14)
$$

where the $\tau_s^{\,\alpha}$ and $\xi_s^{\,k}$ are the generators of the transformation and are given by (4.4).  □

We shall postpone the proof of Theorem 4.1, which is slightly more involved than the proof for the single integral case, until the next section. For now we shall be content to comment on the importance of the theorem and show how it provides easy access to the classical theorem of Emmy Noether.

Identities (4.14) can be interpreted in two ways. If the transformations and an invariant fundamental integral is known, then Eqs. (4.14) represent a set of identities in $t$, the $x^k$, and the derivatives $\dot{x}_\alpha^k$. On the other hand, if the Lagrangian $L$ is unknown, we can think of Eqs. (4.14) as representing $r$ first-order quasi-linear partial differential equations for $L$; consequently, Eqs. (4.14) can serve to characterize the Lagrangians, or fundamental integrals, that possess given invariance properties. Another fundamental importance of Theorem 4.1 is that it leads to a direct proof of Noether's theorem for multiple integrals. The latter is a basic cornerstone of modern field theories in that it provides a direct method for writing down explicitly the conservation laws for the field, given only the invariance properties of the fundamental integral. We now state and prove this theorem.

**4.2 Theorem** (*Noether*) A necessary condition for the fundamental integral $J$ defined by (4.1) to be absolutely invariant under the $r$-parameter family of transformations (4.2) is that the following $r$ identities hold true:

$$-E_k(\xi_s^k - \dot{x}_\alpha^k \tau_s^\alpha) = \frac{\partial}{\partial t^\alpha}\left[\left(L\delta_\beta^\alpha - \dot{x}_\beta^k \frac{\partial L}{\partial \dot{x}_\alpha^k}\right)\tau_s^\beta + \frac{\partial L}{\partial \dot{x}_\alpha^k}\xi_s^k\right]$$

$$(s = 1, \ldots, r), \qquad (4.15)$$

where

$$E_k \equiv \frac{\partial L}{\partial x^k} - \frac{\partial}{\partial t^\alpha}\frac{\partial L}{\partial \dot{x}_\alpha^k}$$

are the Euler–Lagrange expressions.

*Proof* The proof of Theorem 4.2 is similar to that of Theorem 2.2. We note, either by the chain rule for partial differentiation or by the Leibniz rule for derivatives of products, that the following three sets of identities are valid:

$$\frac{\partial L}{\partial t^\alpha}\tau_s^\alpha = \frac{dL}{dt^\alpha}\tau_s^\alpha - \frac{\partial L}{\partial x^k}\dot{x}_\alpha^k\tau_s^\alpha - \frac{\partial L}{\partial \dot{x}_\beta^k}\ddot{x}_{\alpha\beta}^k\tau_s^\alpha,$$

$$\frac{\partial L}{\partial \dot{x}_\alpha^k}\frac{d\xi_s^k}{dt^\alpha} = \frac{d}{dt^\alpha}\left(\frac{\partial L}{\partial \dot{x}_\alpha^k}\xi_s^k\right) - \frac{d}{dt^\alpha}\left(\frac{\partial L}{\partial \dot{x}_\alpha^k}\right)\xi_s^k,$$

$$\frac{\partial L}{\partial \dot{x}_\alpha^k}\dot{x}_\beta^k\frac{d\tau_s^\beta}{dt^\alpha} + \frac{\partial L}{\partial \dot{x}_\alpha^k}\ddot{x}_{\alpha\beta}^k\tau_s^\beta = \frac{d}{dt^\alpha}\left(\frac{\partial L}{\partial \dot{x}_\alpha^k}\dot{x}_\beta^k\tau_s^\beta\right) - \frac{d}{dt^\alpha}\left(\frac{\partial L}{\partial \dot{x}_\alpha^k}\right)\dot{x}_\beta^k\tau_s^\beta.$$

Upon substituting these relations into Eqs. (4.14), we obtain, after simplification, the expressions

$$\frac{d}{dt^\alpha}\left(L\tau_s{}^\alpha + \frac{\partial L}{\partial \dot{x}_\alpha{}^k}\zeta_s{}^k - \frac{\partial L}{\partial \dot{x}_\alpha{}^k}\dot{x}_\beta{}^k\tau_s{}^\beta\right) + \left(\frac{\partial L}{\partial x^k} - \frac{\partial}{\partial t^\alpha}\frac{\partial L}{\partial \dot{x}_\alpha{}^k}\right)(\zeta_s{}^k - \dot{x}_\beta{}^k\tau_s{}^\beta) = 0.$$

In terms of the Euler–Lagrange expressions $E_k$ the last equations yield the Noether identities (4.15), and this completes the proof. □

We remark at this point that the above definitions and theorems have only involved the concept of absolute invariance and not the more general concept of divergence-invariance, as in Chapter 2. The reason for this omission is that we do not present in the multiple integral case any important, or otherwise, application of the latter concept. Therefore, for multiple integrals, we shall drop the word "absolute" and just say "invariant."

## 4.3   DERIVATION OF THE INVARIANCE IDENTITIES

In this section we prove Theorem 4.2 and derive the fundamental invariance identities (4.14). The proof is along the same lines as the proof for the single integral case (Theorem 2.1). The reader who is prepared to accept identities (4.14), or the reader who is not interested in the details of the proof, should skip this section and go directly to Section 4.4 for a discussion of conservation laws.

Before proceeding with the proof, we shall require several little facts which can be obtained easily from the invariance transformation (4.2) [or (4.3)]. We shall list these without proof:

$$\left(\frac{\partial \bar{x}^k}{\partial t^\alpha}\right)_0 = \left(\frac{\partial \bar{t}^\alpha}{\partial x^j}\right)_0 = 0, \qquad \left(\frac{\partial \bar{x}^k}{\partial x^h}\right)_0 = \delta_h{}^k, \qquad \left(\frac{\partial \bar{t}^\alpha}{\partial t^\beta}\right)_0 = \delta_\beta{}^\alpha,$$

$$\left(\frac{\partial^2 \bar{x}^k}{\partial \varepsilon^s\, \partial t^\alpha}\right)_0 = \frac{\partial \zeta_s{}^k}{\partial t^\alpha}, \qquad \left(\frac{\partial^2 \bar{x}^k}{\partial \varepsilon^s\, \partial x^h}\right)_0 = \frac{\partial \zeta_s{}^k}{\partial x^h}, \tag{4.16}$$

$$\left(\frac{\partial^2 \bar{t}^\alpha}{\partial \varepsilon^s\, \partial t^\beta}\right)_0 = \frac{\partial \tau_s{}^\alpha}{\partial t^\beta}, \qquad \left(\frac{\partial^2 \bar{t}^\alpha}{\partial \varepsilon^s\, \partial x^h}\right)_0 = \frac{\partial \tau_s{}^\alpha}{\partial x^k},$$

where $(\cdot)_0$ means $(\cdot)_{\varepsilon=0}$, and $\delta_\beta{}^\alpha$ is the Kronecker delta.

As in the case of Theorem 2.1, we differentiate the expression which defines the invariance (in this case (4.7)) with respect to $\varepsilon^s$, after which we put $\varepsilon = 0$. Our argument here closely follows the proof given by Rund in [4], except for some reorganization and the inclusion of more detail.

Differentiating (4.7) with respect to $\varepsilon^s$ and setting $\varepsilon = 0$ gives, after applying the chain rule for derivatives and Eqs. (4.16),

$$\left\{\frac{\partial L}{\partial t^\alpha}\,\tau_s^{\ \alpha} + \frac{\partial L}{\partial x^k}\,\xi_s^{\ k} + \frac{\partial L}{\partial \dot{x}_\alpha^{\ k}}\left(\frac{\partial \dot{\bar{x}}_\alpha^{\ k}}{\partial \varepsilon^s}\right)_0\right\}\left[\det\!\left(\frac{\partial \bar{t}^\alpha}{\partial t^\beta}\right)\right]_0 + L\left[\frac{\partial}{\partial \varepsilon^s}\det\!\left(\frac{\partial \bar{t}^\alpha}{\partial t^\beta}\right)\right]_0$$

$$= \frac{\partial}{\partial t^\alpha}\,\Phi_s^{\ \alpha}, \tag{4.17}$$

where $\partial \dot{\bar{x}}_\alpha^{\ k}/\partial \varepsilon^s$ denotes $(\partial/\partial \varepsilon^s)(\partial \bar{x}^k/\partial \bar{t}^\alpha)$. At this point it is necessary to calculate the terms

$$\left(\frac{\partial \dot{\bar{x}}_\alpha^{\ k}}{\partial \varepsilon^s}\right)_0, \quad \left[\det\!\left(\frac{\partial \bar{t}^\alpha}{\partial t^\beta}\right)\right]_0, \quad \text{and} \quad \left[\frac{\partial}{\partial \varepsilon^s}\det\!\left(\frac{\partial \bar{t}^\alpha}{\partial t^\beta}\right)\right]_0.$$

First, it is obvious that

$$\left[\det\!\left(\frac{\partial \bar{t}^\alpha}{\partial t^\beta}\right)\right]_0 = 1. \tag{4.18}$$

Next, in order to compute the last term above, it is necessary to recall the rule for differentiating a determinant: if $a = \det(a_j^{\ i})$, where the $a_j^{\ i}$'s are functions of some parameter $\alpha$, then $\partial a/\partial \alpha = (\partial a_j^{\ i}/\partial \alpha)A_i^{\ j}$ (sum on $i$, $j$), where $A_i^{\ j}$ is the cofactor of $a_j^{\ i}$ in the determinant. Applying this rule to $\det(\partial \bar{t}^\alpha/\partial t^\beta)$, we obtain

$$\frac{\partial}{\partial \varepsilon^s}\det\!\left(\frac{\partial \bar{t}^\alpha}{\partial t^\beta}\right) = \frac{\partial}{\partial \varepsilon^s}\left(\frac{\partial \bar{t}^\alpha}{\partial t^\beta}\right)A_\alpha^{\ \beta}, \tag{4.19}$$

where $A_\alpha^{\ \beta}$ is the cofactor of the element $\partial \bar{t}^\alpha/\partial t^\beta$. Applying the chain rule for derivatives to (4.19), we conclude that

$$\frac{\partial}{\partial \varepsilon^s}\det\!\left(\frac{\partial \bar{t}^\alpha}{\partial t^\beta}\right) = \left(\frac{\partial^2 \bar{t}^\alpha}{\partial \varepsilon^s\,\partial t^\beta} + \frac{\partial^2 \bar{t}^\alpha}{\partial \varepsilon^s\,\partial x^k}\,\dot{x}_\beta^{\ k}\right)A_\alpha^{\ \beta}.$$

Upon evaluation of this equation at $\varepsilon = 0$, we have, using (4.16),

$$\left[\frac{\partial}{\partial \varepsilon^s}\det\!\left(\frac{\partial \bar{t}^\alpha}{\partial t^\beta}\right)\right]_0 = \left(\frac{\partial \tau_s^{\ \alpha}}{\partial t^\beta} + \frac{\partial \tau_s^{\ \alpha}}{\partial x^k}\,\dot{x}_\beta^{\ k}\right)(A_\alpha^{\ \beta})_0.$$

Noting that

$$(A_\alpha^{\ \beta})_0 = \delta_\alpha^{\ \beta},$$

we finally obtain

$$\left[\frac{\partial}{\partial \varepsilon^s}\det\!\left(\frac{\partial \bar{t}^\alpha}{\partial t^\beta}\right)\right]_0 = \frac{d\tau_s^{\ \alpha}}{dt^\beta}\,\delta_\alpha^{\ \beta} = \frac{d\tau_s^{\ \alpha}}{dt^\alpha}. \tag{4.20}$$

The calculation of $(\partial \bar{\dot{x}}_\alpha{}^k / \partial \varepsilon^s)_0$ is a little more involved. We begin by differentiating the equation $\bar{x}^k = \bar{x}^k(\bar{t})$ with respect to $t^\beta$ to obtain

$$\frac{\partial \bar{x}^k}{\partial t^\beta} + \frac{\partial \bar{x}^k}{\partial x^h} \dot{x}_\beta{}^h = \frac{\partial \bar{x}^k}{\partial \bar{t}^\alpha} \frac{d\bar{t}^\alpha}{dt^\beta}$$

$$= \frac{\partial \bar{x}^k}{\partial \bar{t}^\alpha} \left( \frac{\partial \bar{t}^\alpha}{\partial t^\beta} + \frac{\partial \bar{t}^\alpha}{\partial x^h} \dot{x}_\beta{}^h \right). \tag{4.21}$$

Putting $\varepsilon = 0$ we find that

$$\delta_h{}^k \dot{x}_\beta{}^h = \left( \frac{\partial \bar{x}^k}{\partial \bar{t}^\alpha} \right)_0 \delta_\beta{}^\alpha$$

or

$$\dot{x}_\beta{}^k = \left( \frac{\partial \bar{x}^k}{\partial \bar{t}^\beta} \right)_0, \tag{4.22}$$

where we have made free use of Eqs. (4.16). Now, differentiation of (4.21) with respect to $\varepsilon^s$ gives

$$\frac{\partial^2 \bar{x}^k}{\partial \varepsilon^s \, \partial t^\beta} + \frac{\partial^2 \bar{x}^k}{\partial \varepsilon^s \, \partial x^h} \dot{x}_\beta{}^h = \frac{\partial \bar{x}^k}{\partial \bar{t}^\alpha} \left( \frac{\partial^2 \bar{t}^\alpha}{\partial \varepsilon^s \, \partial t^\beta} + \frac{\partial^2 \bar{t}^\alpha}{\partial \varepsilon^s \, \partial x^h} \dot{x}_\beta{}^h \right)$$

$$+ \frac{\partial}{\partial \varepsilon^s} \frac{\partial \bar{x}^k}{\partial \bar{t}^\alpha} \left( \frac{\partial \bar{t}^\alpha}{\partial t^\beta} + \frac{\partial \bar{t}^\alpha}{\partial x^h} \dot{x}_\beta{}^h \right).$$

Putting $\varepsilon = 0$ and again using (4.16), we get

$$\frac{\partial \xi_s{}^k}{\partial t^\beta} + \frac{\partial \xi_s{}^k}{\partial x^h} \dot{x}_\beta{}^h = \dot{x}_\alpha{}^k \left( \frac{\partial \tau_s{}^\alpha}{\partial t^\beta} + \frac{\partial \tau_s{}^\alpha}{\partial x^h} \dot{x}_\beta{}^h \right) + \left( \frac{\partial}{\partial \varepsilon^s} \bar{\dot{x}}_\alpha{}^k \right)_0 \delta_\beta{}^\alpha, \tag{4.23}$$

where we have also used (4.22). Writing (4.23) in terms of the total derivatives of the generators, we conclude that

$$\left( \frac{\partial}{\partial \varepsilon^s} \bar{\dot{x}}_\beta{}^k \right)_0 = \frac{d\xi_s{}^k}{dt^\beta} - \dot{x}_\alpha{}^k \frac{d\tau_s{}^\alpha}{dt^\beta}. \tag{4.24}$$

Finally, substitution of (4.24), (4.20), and (4.18) into (4.17) yields the invariance identities (4.14); this, then, completes the proof of Theorem (4.1).  □

## 4.4  CONSERVATION THEOREMS

We have seen in the single integral case that Noether's theorem led to first integrals of the governing differential equations, the Euler–Lagrange equations. For multiple integrals the conservation laws take the form of a vanishing divergence, as the following corollary of Theorem 4.2 shows.

**4.1 Corollary** If the fundamental integral $J$ defined by (4.1) is absolutely invariant under the $r$-parameter family of transformations given by (4.2), and if $E_k = 0$ for $k = 1, \ldots, n$, then the following $r$ identities hold true:

$$\frac{\partial}{\partial t^\alpha} \left\{ \left( L\delta_\beta{}^\alpha - \dot{x}_\beta{}^k \frac{\partial L}{\partial \dot{x}_\alpha{}^k} \right) \tau_s{}^\beta + \frac{\partial L}{\partial \dot{x}_\alpha{}^k} \xi_s{}^k \right\} = 0 \qquad (s = 1, \ldots, r). \quad (4.25)$$

The proof is an immediate consequence of Theorem 4.2. $\square$

Let us define the $m \times r$ quantities

$$\psi_s{}^\alpha \equiv \left( L\delta_\beta{}^\alpha - \dot{x}_\beta{}^k \frac{\partial L}{\partial \dot{x}_\alpha{}^k} \right) \tau_s{}^\beta + \frac{\partial L}{\partial \dot{x}_\alpha{}^k} \xi_s{}^k. \quad (4.26)$$

Then (4.25) states that

$$\frac{\partial \psi_s{}^\alpha}{\partial t^\alpha} = 0 \qquad (s = 1, \ldots, r), \quad (4.27)$$

each equation of which is the form of an equation of continuity and is interpreted as a conservation law. We witness, for example, in electrodynamics, the usual statement of conservation of charge:

$$\frac{\partial \rho}{\partial t} + \operatorname{div} \mathbf{J} = 0,$$

where $\rho$ is the charge density and $\mathbf{J} = (J^2, J^3, J^4)$ is the current density. If we define a four-vector $(J^1, J^2, J^3, J^4)$, where $J^1 = \rho$, then the conservation of charge law is given by

$$\frac{\partial J^\alpha}{\partial t^\alpha} = 0 \qquad (\text{sum on } \alpha = 1, 2, 3, 4), \quad (4.28)$$

where $(t^1, t^2, t^3, t^4)$ are the coordinates of time–space. Equation (4.28) has obviously the same form as Eq. (4.27).

We now explore a general interpretation of Eqs. (4.27). Let us single out in $(t^1, \ldots, t^m)$ the variable $t^1$, which can be thought of as representing time, the remaining variables representing spatial coordinates. Now, let $V$ be the hypercylinder in $R^m$ defined by $V = [t_1, t_2] \times S$, where $S$ is the "sphere"

$$(t^2)^2 + (t^3)^2 + \cdots + (t^m)^2 \le c^2$$

(see Fig. 6 for a geometrical picture when $m = 3$). Integrating (4.27) over $V$ and applying the Divergence Theorem of Gauss, we obtain

$$0 = \int_V \frac{\partial \psi_s{}^\alpha}{\partial t^\alpha} \, dt^1 \cdots dt^m = \int_{\partial V} \psi_s{}^\alpha n_\alpha \, d\sigma, \quad (4.29)$$

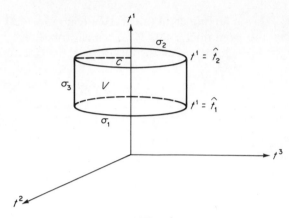

**FIGURE 6**

where $\partial V$ denotes the boundary of $V$ and $n_\alpha$ denote the components of the outer normal to $V$. We note that the boundary of $V$ consists of three portions. These are

(i)  $\sigma_1$ and $\sigma_2$, the hyperplanes $t^1 = \hat{t}_1$ and $t^1 = \hat{t}_2$ which are cut off by the hypercylinder $(t^2)^2 + \cdots + (t^m)^2 = c^2$.

(ii)  $\sigma_3$, the lateral side of the hypercylinder defined by

$$(t^2)^2 + \cdots + (t^m)^2 = c^2, \qquad \hat{t}_1 \le t^1 \le \hat{t}_2.$$

Therefore, from (4.29), we can write

$$\int_{\sigma_1} \psi_s{}^\alpha n_\alpha \, d\sigma + \int_{\sigma_2} \psi_s{}^\alpha n_\alpha \, d\sigma + \int_{\sigma_3} \psi_s{}^\alpha n_\alpha \, d\sigma = 0. \qquad (4.30)$$

We *assume* (in accord with the physical fact that the field vanishes at infinity) that

$$\lim_{c \to \infty} \int_{\sigma_3} \psi_s{}^\alpha n_\alpha \, d\sigma = 0, \qquad (s = 1, \ldots, r).$$

Upon taking the limit as $c \to \infty$ in (4.30), we observe that this assumption leads to

$$\int_{\hat{\sigma}_1} \psi_s{}^\alpha n_\alpha \, d\sigma + \int_{\hat{\sigma}_2} \psi_s{}^\alpha n_\alpha \, d\sigma = 0, \qquad (4.31)$$

where $\hat{\sigma}_1$ and $\hat{\sigma}_2$ are the infinite hyperplanes $t^1 = \hat{t}_1$ and $t^1 = \hat{t}_2$. But the outer oriented normals on these two hyperplanes are given, respectively, by $n_1 = -1$, $n_2 = \cdots = n_m = 0$ and $n_1 = 1$, $n_2 = \cdots = n_m = 0$. Consequently, (4.31) becomes

$$- \int_{\hat{\sigma}_1} \psi_s{}^1 \, dt^2 \cdots dt^m + \int_{\hat{\sigma}_2} \psi_s{}^1 \, dt^2 \cdots dt^m = 0,$$

where we have noted that $d\sigma = dt^2 \cdots dt^m$. In other words,

$$\int_{t^1 = \hat{t}_1} \psi_s{}^1 \, dt^2 \cdots dt^m = \int_{t^1 = \hat{t}_2} \psi_s{}^1 \, dt^2 \cdots dt^m. \tag{4.32}$$

Since $\hat{t}_1$ and $\hat{t}_2$ are arbitrary, it follows from (4.32) that the $r$ expressions

$$\int_{t^1 = \text{constant}} \psi_s{}^1 \, dt^2 \cdots dt^m \qquad (s = 1, \ldots, r) \tag{4.33}$$

are independent of $t^1$ and hence represent *conserved quantities* for the system satisfying Eqs. (4.27).

In terms of the so-called canonical variables for the multiple integral problem it is possible to give a concise statement of the conservation laws (4.25). Let us define the *canonical momenta* by

$$p_k{}^\alpha \equiv \frac{\partial L}{\partial \dot{x}_\alpha{}^k} \tag{4.34}$$

and the *Hamiltonian complex* by

$$H_\beta{}^\alpha \equiv -L\delta_\beta{}^\alpha + \dot{x}_\beta{}^k \frac{\partial L}{\partial \dot{x}_\alpha{}^k}. \tag{4.35}$$

Then (4.25) becomes

$$\frac{\partial}{\partial t^\alpha} (-H_\beta{}^\alpha \tau_s{}^\beta + p_k{}^\alpha \xi_s{}^k) = 0 \qquad (s = 1, \ldots, r). \tag{4.36}$$

**4.2  Example**  To illustrate Corollary 4.1, we assume that fundamental integral (4.1) is absolutely invariant under the $m$-parameter family of transformations

$$\bar{t}^\alpha = t^\alpha + \varepsilon^\alpha, \qquad \bar{x}^k = x^k,$$

which represents a translation of $(t^1, \ldots, t^m)$-space. The generators of this family are given by

$$\tau_s{}^\alpha = \delta_s{}^\alpha, \qquad \xi_s{}^k = 0 \qquad (\alpha, s = 1, \ldots, m).$$

Therefore, the conservation laws (4.36) become

$$\frac{\partial}{\partial t^\alpha}(-H_\beta{}^\alpha\delta_s{}^\beta) = 0$$

or

$$\frac{\partial H_s{}^\alpha}{\partial t^\alpha} = 0 \qquad (s = 1, \ldots, m).$$

Consequently, via (4.33), the conserved quantities are

$$\int_{t^1 = \text{constant}} H_s{}^1 \, dt^2 \cdots dt^m \qquad (s = 1, \ldots, m).$$

## EXERCISES

**4-1**  Consider the functional

$$J(\phi) = \iint_R K(s, t)\phi(s)\phi(t) \, ds \, dt + \int_a^b \phi^2(s) \, ds - 2 \int_a^b \phi(s) f(s) \, ds$$

for $\phi \in C[a, b]$, where $K(s, t)$ is a given continuous, symmetric function of $s$ and $t$ on the square $R = \{(s, t) | a \le s, t \le b\}$, and $f$ is a given function of class $C[a, b]$. Compute the first variation of $J$ and obtain a necessary condition for an extremum.

**4-2**  Let $G \subset R^m$ be a closed, bounded set and let $U \subset R^r$ be an open neighborhood about the origin in $R^r$. Let $\gamma: G \times U \to R^m$ be a given twice continuously differentiable function satisfying $\gamma(t, 0) = t$ for all $t \in G$. Prove that there exists a $d > 0$ such that for any $\varepsilon \in U$ and $|\varepsilon| < d$, the function $\gamma_\varepsilon: G \to R^m$ defined by $\gamma_\varepsilon(t) = \gamma(t, \varepsilon)$ has a twice continuously differentiable inverse. (Note that this exercise is the multidimensional analog of Lemma 2.1.)

[*Hint*: Show that the Jacobian is nonzero and use compactness and the inverse function theorem to conclude that $\gamma_\varepsilon$ is locally invertible. Using this fact, show that $\gamma_\varepsilon$ has a global inverse.]

**4-3**  Referring to Example 4.1, use Noether's theorem to explicitly write down the conservation laws for this problem.

**4-4**  Write the Korteweg–deVries equation $u_t - 6uu_x + u_{xxx} = 0$ in the form of a conservation law.

    *Answer*:  $u_t + (-3u^2 + u_{xx})_x = 0.$

**4-5** Directly from the Korteweg–deVries equation in Exercise 4-4 derive the conservation law

$$\frac{\partial}{\partial t} u^2 + \frac{\partial}{\partial x} (2uu_{xx} - 4u^3 - u_x{}^2) = 0.$$

[*Hint*: Multiply the Korteweg–deVries equation by $2u$.]

**4-6** Directly from the Korteweg–deVries equation in Exercise 4-4 derive the conservation law

$$\frac{\partial}{\partial t} (u^3 + \tfrac{1}{2}u_x{}^2) + \frac{\partial}{\partial x} (-\tfrac{9}{2}u^4 + 3u^2 u_{xx} - 6uu_x{}^2 + u_x u_{xxx} - \tfrac{1}{2}u_{xx}^2) = 0.$$

[*Hint*: Multiply by $3u^2 - u_{xx}$ and use the Korteweg–deVries equation to eliminate derivatives of $u$ with respect to $t$.]

# CHAPTER 5

# Invariance Principles
# in the Theory of Physical Fields

## 5.1  INTRODUCTION

Invariance considerations in field theory arise naturally from the postulates of the theory of special relativity. As we shall observe, those postulates force upon us the requirement that any physical field, and in particular the variational principle for that field, must be invariant under Lorentz transformations. Therefore we shall obtain, a priori, a family of transformations under which the action integral will be invariant and from which a direct application of the Noether theorem will yield conservation laws for the field.

The special theory of relativity had its origin in electromagnetism in the last part of the nineteenth century. At that time it was well known that the laws of classical mechanics were the same in all coordinate systems moving uniformly relative to each other (*inertial frames*); that is to say, the laws of mechanics are invariant under the so-called *Galilean transformations*

$$\bar{t} = t,$$
$$\bar{x} = x + v_1 t,$$
$$\bar{y} = y + v_2 t,$$
$$\bar{z} = z + v_3 t,$$

where $v = (v_1, v_2, v_3)$ is the relative velocity between two inertial frames $(t, x, y, z)$ and $(\bar{t}, \bar{x}, \bar{y}, \bar{z})$. This invariance was explicitly demonstrated in Chapter 2 in the case of $n$ particles interacting under the inverse square law. However, it was discovered that the fundamental laws of electromagnetism, namely, Maxwell's equations, did not possess this Galilean invariance. Therefore, it appeared that either Maxwell's equations were not correct, the correct equations being Galilean invariant and not yet known, or that Galilean invariance applied only to classical mechanics, the conclusion being that electromagnetism has its own preferred inertial frame, the one in which the medium of propagation (the "ether") was at rest. Experiments seemed to indicate, however, that neither of these possibilities were tenable. Then Lorentz, in an effort to reconcile the fact that electromagnetic theory failed to be Galilean invariant, showed in 1904 that Maxwell's equations *in vacuo* were invariant under the so-called Lorentz transformations. After important contributions from Minkowski and Poincaré, it remained for Einstein to consider another possibility, namely, that there existed a principle of relativity different from Galilean relativity which would govern both classical mechanics and electromagnetism; this would imply, of course, that the laws of mechanics were not, as they stood, correct. This assumption led him in 1905 to formulate the special theory of relativity which is at the present time believed to encompass all types of interactions, with the single exception of large scale gravitational phenomena. It is this theory, based on the Lorentz transformations, that is the foundation of invariance requirements for variational principles in the theory of physical fields.

Before embarking upon a detailed discussion of these notions, we shall require two introductory topics. We shall begin in Section 2 with a short introduction to tensor analysis followed by a section on the Lorentz group. After that we shall formulate the invariance criteria for fields, in general. Our program will continue with a discussion of scalar and covariant vector fields, finally concluding with a detailed application to the electromagnetic field.

## 5.2 TENSORS

By a physical field we roughly mean a function, or collection of functions, which describes some physical quantity or quantities at all points in some specified region of space–time. When the field can be described by a single real-valued function, i.e., when the physical quantity can be completely specified by a single real number at each point independent of any coordinate system, then the field is called a *scalar field*. Examples are temperature, density, and electrostatic potential. A *vector field*, on the other hand, requires at each point an $n$-tuple of real numbers to specify the given quantity; viewed over an extended region of space–time, a vector field is an $n$-tuple of functions, each defined on the given region. Examples are velocity, force, and the electromagnetic field intensity. For example, the electric field intensity $\mathbf{E}(x, y, z) = (E_1(x, y, z), E_2(x, y, z), E_3(x, y, z))$ is a vector field which requires the three functions $E_i$, the so-called components of the field, for its complete description. Geometrically, we generally think of a vector (at least in three dimensions) as an ordered triple of numbers which has both magnitude and direction, whereas a scalar has only magnitude. However, this picture is not completely discriptive, but rather some notion of how the components are altered when the coordinates are changed is also required. In fact, this latter idea is precisely the one that is involved in the definition of a tensor, of which vectors and scalars are special cases. In loose terms, a tensor is a multi-indexed set of numbers which transform in a specified manner under change of coordinates. In the following, we shall restrict the discussion to four dimensions, having in mind the applications to come; no generality is lost in this restriction.

More precisely, let $X_4$ be a four-dimensional *manifold*, i.e., a set of points which is locally a Euclidean space in such a manner that two different coordinatizations $(t^0, t^1, t^2, t^3)$ and $(\bar{t}^0, \bar{t}^1, \bar{t}^2, \bar{t}^3)$ of a neighborhood about a point $P$ are connected through continuously differentiable, one-to-one, invertible mappings

$$\bar{t}^\alpha = f^\alpha(t^0, t^1, t^2, t^3) \tag{5.1}$$

and

$$t^\alpha = h^\alpha(\bar{t}^0, \bar{t}^1, \bar{t}^2, \bar{t}^3) \qquad (\alpha = 0, 1, 2, 3). \tag{5.2}$$

The $4 \times 4$ matrix with elements $(\partial \bar{t}^\alpha / \partial t^\beta)$ is called the *Jacobi matrix* of the transformation; it is, by our assumptions, nonsingular and its inverse is the matrix with elements $(\partial t^\alpha / \partial \bar{t}^\beta)$. That is,

$$\frac{\partial \bar{t}^\alpha}{\partial t^\beta} \frac{\partial t^\beta}{\partial \bar{t}^\gamma} = \delta_\gamma{}^\alpha.$$

Here, as in earlier chapters, we are using the summation convention of summing over the repeated index $\beta$ over its range, in this case from 0 to 3. The determinant of the Jacobi matrix is called the *Jacobian* of the transformation. If we denote it by $J$, then

$$J = \det\left(\frac{\partial \bar{t}^\alpha}{\partial t^\beta}\right).$$

Now, a set of $4^{m+n}$ indexed numbers $T^{\alpha_1 \cdots \alpha_m}{}_{\beta_1 \cdots \beta_n}$ is said to form the components of a *tensor of type* $\binom{m}{n}$ *and rank* $m + n$ if they transform according to the rule

$$\bar{T}^{\alpha_1 \cdots \alpha_m}{}_{\beta_1 \cdots \beta_n} = \frac{\partial t^{\eta_1}}{\partial \bar{t}^{\beta_1}} \cdots \frac{\partial t^{\eta_n}}{\partial \bar{t}^{\beta_n}} \frac{\partial \bar{t}^{\alpha_1}}{\partial t^{\gamma_1}} \cdots \frac{\partial \bar{t}^{\alpha_m}}{\partial t^{\gamma_m}} T^{\gamma_1 \cdots \gamma_m}{}_{\eta_1 \cdots \eta_n} \tag{5.3}$$

when passing from the $(t^\alpha)$ system to the $(\bar{t}^\alpha)$ system; the bar over the $T$ denotes, of course, the components of the tensor in the barred coordinate system, and the summation convention is implicitly assumed. The nonnegative integers $m$ and $n$ are said to be the degree of *contravariance* and *covariance* of the tensor, respectively. Equation (5.3) is called the *transformation law* for the tensor, and it is assumed that all derivatives which occur in (5.3) are evaluated at the point $P$. Taking a specific example, a tensor of type $\binom{2}{1}$, or twice contravariant and once covariant, is in a given $(t^\alpha)$ coordinate system a set of $4^3 = 64$ real numbers $T^{\alpha\beta}{}_\gamma$ which transform to the $(\bar{t}^\alpha)$ system according to the transformation law

$$\bar{T}^{\alpha\beta}{}_\gamma = \frac{\partial t^\mu}{\partial \bar{t}^\gamma} \frac{\partial \bar{t}^\alpha}{\partial \bar{t}^\nu} \frac{\partial \bar{t}^\beta}{\partial t^\lambda} T^{\nu\lambda}{}_\mu.$$

The definition that we have presented defines a tensor in terms of its components; i.e., a tensor is a set of components which transforms in a specified manner when passing to a new coordinate system. An alternate approach with a little more mathematical flavor reveals a different, but equivalent, nature of the notion of a tensor. Such an approach, which we shall only briefly mention, emphasizes the algebraic structure of a tensor rather than, as we have defined it, its transformation properties. At the point $P$ in $X_4$ it can be easily shown that there is attached a linear space (the so-called tangent vector space) $V_p$ of dimension four. As a result, we can form the linear space $V_p^*$ dual to $V_p$ consisting of all the linear functionals on $V$. A tensor $T$ of type $\binom{m}{n}$ can then be defined as a multilinear mapping from the cartesian product

$$\underbrace{V_p^* \times \cdots \times V_p^*}_{m \text{ times}} \times \underbrace{V_p \times \cdots \times V_p}_{n \text{ times}}$$

into the real numbers. If $e_i$ is a basis for $V_p$ and $\varepsilon^j$ is the corresponding natural basis for the dual space $V_p^*$ defined by $\varepsilon^j(e_i) = \delta_i{}^j$, where $\delta_i{}^j$ is the Kronecker delta, then the components of $T$ are given by

$$T^{\alpha_1 \cdots \alpha_m}{}_{\beta_1 \cdots \beta_n} = T(\varepsilon^{\alpha_1}, \ldots, \varepsilon^{\alpha_m}, e_{\beta_1}, \ldots, e_{\beta_n}).$$

That is, the components of a tensor $T$ are just the numbers obtained by evaluating $T$ on the corresponding basis vectors. The transformation law follows easily from the linearity of $T$. For, suppose $\bar{e}_i$ is another basis of $V_p$ and $\bar{\varepsilon}^j$ is the corresponding dual basis. These basis vectors are related to the original basis vectors via linear transformations

$$\bar{e}_i = a_i{}^j e_j \quad \text{and} \quad \bar{\varepsilon}^i = b_j{}^i \varepsilon^j.$$

If we restrict our discussion, for simplicity, to a tensor $T: V_p^* \times V_p^* \times V_p \to R^1$ of type $\binom{2}{1}$, then the components in the barred system are

$$\begin{aligned}
\bar{T}^{ij}{}_k &= T(\bar{\varepsilon}^i, \bar{\varepsilon}^j, e_k) \\
&= T(b_p{}^i \varepsilon^p, b_q{}^j \varepsilon^q, a_k{}^r e_r) \\
&= b_p{}^i b_q{}^j a_k{}^r T(\varepsilon^p, \varepsilon^q, e_r) \\
&= b_p{}^i b_q{}^j a_k{}^r T^{pq}{}_r,
\end{aligned}$$

which is the general transformation law of the components of the tensor $T$ of type $\binom{2}{1}$. When the bases of $V_p$ and $V_p^*$ are obtained from the coordinate vector fields with respect to two coordinate systems $(t^i)$ and $(\bar{t}^i)$ at $P$, i.e., $e_i = \partial/\partial t^i$, $\bar{e}_i = \partial/\partial \bar{t}^i$, $\varepsilon^i = dt^i$, and $\bar{\varepsilon}^i = d\bar{t}^i$, then $a_j{}^i = \partial t^i/\partial \bar{t}^j$ and $b_j{}^i = \partial \bar{t}^i/\partial t^j$ and the general transformation law given above reduces to

$$\bar{T}^{ij}{}_k = \frac{\partial \bar{t}^i}{\partial t^p} \frac{\partial \bar{t}^j}{\partial t^q} \frac{\partial t^r}{\partial \bar{t}^k} T^{pq}{}_r,$$

which is the classical statement of the transformation law for tensors of type $\binom{2}{1}$. A detailed treatment of this approach can be found in Bishop and Goldberg [1].

As hinted in the last paragraph, the tensors of a given type at $P$ have the algebraic structure of a linear space; i.e., they can be added and multiplied by real numbers with the usual axioms holding true. For example, if $T^{\alpha\beta}{}_\gamma$ and $U^{\alpha\beta}{}_\gamma$ are two tensors at $P$, then the *sum*

$$S^{\alpha\beta}{}_\gamma \equiv T^{\alpha\beta}{}_\gamma + U^{\alpha\beta}{}_\gamma$$

is a tensor of the same type, and so is the scalar multiple

$$M^{\alpha\beta}{}_\gamma \equiv r T^{\alpha\beta}{}_\gamma, \qquad r \in R^1.$$

To show that $S^{\alpha\beta}{}_\gamma$ and $M^{\alpha\beta}{}_\gamma$ are tensors, we must show that they satisfy the appropriate transformation law; this is an easy exercise.

Tensors of different type can be multiplied componentwise to form new tensors of higher rank. For example, let $A^\alpha$ and $B^\alpha{}_\beta$ be two tensors of type $\binom{1}{0}$ and $\binom{1}{1}$, respectively, at $P$. Then we define the *product*

$$C^{\alpha\beta}{}_\gamma \equiv A^\alpha B^\beta{}_\gamma.$$

It is easy to see that the $C^{\alpha\beta}{}_\gamma$ are the components of a tensor of type $\binom{2}{1}$. We will present this simple argument since it is representative of similar arguments to show that a given set of components form a tensor. Since $A^\alpha$ and $B^\beta{}_\gamma$ are tensors, they transform according to

$$\bar{A}^\alpha = \frac{\partial \bar{t}^\alpha}{\partial t^\nu} A^\nu, \qquad \bar{B}^\beta{}_\gamma = \frac{\partial t^\lambda}{\partial \bar{t}^\gamma} \frac{\partial \bar{t}^\beta}{\partial t^\mu} B^\mu{}_\lambda.$$

Therefore

$$\bar{C}^{\alpha\beta}{}_\gamma = \bar{A}^\alpha \bar{B}^\beta{}_\gamma$$

$$= \frac{\partial \bar{t}^\alpha}{\partial t^\nu} A^\nu \frac{\partial t^\lambda}{\partial \bar{t}^\gamma} \frac{\partial \bar{t}^\beta}{\partial t^\mu} B^\mu{}_\lambda$$

$$= \frac{\partial \bar{t}^\alpha}{\partial t^\nu} \frac{\partial \bar{t}^\beta}{\partial t^\mu} \frac{\partial t^\lambda}{\partial \bar{t}^\gamma} C^{\nu\mu}{}_\lambda.$$

Hence $C^{\alpha\beta}{}_\gamma$ is a tensor since the appropriate transformation law is obeyed.

As it turns out, scalars and vectors are just special types of tensors. A *scalar* (or *invariant*) is a tensor of type $\binom{0}{0}$; hence, it is a single real number which is the same in all coordinate systems. A tensor of type $\binom{1}{0}$ is called a *contravariant vector* and a tensor of type $\binom{0}{1}$ is called a *covariant vector*. Thus, a contravariant vector consists of four components $A^\alpha$ which transform according to

$$\bar{A}^\alpha = \frac{\partial \bar{t}^\alpha}{\partial t^\beta} A^\beta \qquad \text{(contravariant vector)} \tag{5.4}$$

and a covariant vector consists of four components $A_\alpha$ (index down) which transform according to

$$\bar{A}_\alpha = \frac{\partial t^\beta}{\partial \bar{t}^\alpha} A_\beta \qquad \text{(covariant vector).} \tag{5.5}$$

Of course, the above definitions are at the fixed point $P$ of $X_4$.

There is another important way in which to form a new tensor out of a given one. This process, known as *contraction*, involves summing over an upper and lower index. For example, let $T^{\alpha\beta}{}_{\gamma\delta\varepsilon}$ be a tensor of type $\binom{2}{3}$. Then we can define a contracted tensor

$$C^\beta{}_{\gamma\varepsilon} \equiv T^{\alpha\beta}{}_{\gamma\alpha\varepsilon} \qquad \text{(sum on } \alpha\text{)!}$$

Again, it is not difficult to prove that $C^\beta{}_{\gamma\varepsilon}$ is a tensor. To carry out contraction, the given tensor must be of type $\binom{m}{n}$ where $m, n \geq 1$; the contracted tensor is a type $\binom{m-1}{n-1}$. Any upper or lower index may be chosen, but different choices lead to different contracted tensors.

Up to now all of our definitions have been restricted to the point $P$ in the manifold $X_4$. Now we shall introduce the idea of a *tensor field*, which is nothing more than an assignment of a tensor of a given type, say $\binom{m}{n}$, to each point of $X_4$. In terms of a local coordinate system $(t^\alpha)$, we may write, in the coordinate neighborhood, the tensor as a function of the local coordinates; i.e.,†

$$T^{\alpha_1 \cdots \alpha_m}_{\beta_1 \cdots \beta_n} = T^{\alpha_1 \cdots \alpha_m}_{\beta_1 \cdots \beta_n}(t^0, t^1, t^2, t^3).$$

We can then, of course, discuss the differentiability properties of tensors with respect to the $t^\alpha$, etc. Specializing these notions, a *scalar field* is an assignment of a real number, independently of coordinates, to each point of $X_4$. A *contravariant vector field* is a tensor field of type $\binom{1}{0}$ and a *covariant vector field* is a tensor field of type $\binom{0}{1}$.

We shall now introduce some additional structure on $X_4$. We assume that there is given, a priori, a tensor field $g_{\alpha\beta}$ of type $\binom{0}{2}$ which is symmetric in the indices $\alpha$ and $\beta$ (i.e., $g_{\alpha\beta} = g_{\beta\alpha}$). In local coordinates $t^0, \ldots, t^3$, we can write $g_{\alpha\beta} = g_{\alpha\beta}(t^0, t^1, t^2, t^3)$. Our assumption is that in the local coordinate neighborhood, the tensor $g_{\alpha\beta}$ defines the distance $ds^2$ between two nearby points $(t^0, t^1, t^2, t^3)$ and $(t^0 + dt^0, t^1 + dt^1, t^2 + dt^2, t^3 + dt^3)$ according to

$$ds^2 = g_{\alpha\beta}\, dt^\alpha\, dt^\beta. \tag{5.6}$$

For this reason, $g_{\alpha\beta}$ is called the *metric tensor*. For example, if $g_{\alpha\beta} = 0$ ($\alpha \neq \beta$), $g_{\alpha\beta} = 1$ ($\alpha = \beta$), then (5.6) is just a statement of the Pythagorean theorem giving the ordinary Euclidean distance.

Now we define $g^{\alpha\beta}$ by the relation

$$g_{\gamma\beta}g^{\alpha\beta} = \delta_\gamma{}^\alpha, \tag{5.7}$$

where $\delta_\gamma{}^\alpha$ is the Kronecker delta. It is easily checked that $g^{\alpha\beta}$ is a tensor of type $\binom{2}{0}$. With the two tensors $g_{\alpha\beta}$ and its "inverse" $g^{\alpha\beta}$ we can define two other important operations on tensors, namely, *raising and lowering indices*. Given a tensor $T^{\alpha\beta}{}_\gamma$ we raise the index $\gamma$ by defining

$$T^{\alpha\beta\gamma} \equiv T^{\alpha\beta}{}_\varepsilon g^{\varepsilon\gamma}.$$

Or, we can lower the index $\alpha$, say, by defining

$$T_\alpha{}^\beta{}_\gamma \equiv T^{\varepsilon\beta}{}_\gamma g_{\varepsilon\alpha}.$$

We have used the same symbol $T$ because $T^{\alpha\beta}{}_\gamma$, $T^{\alpha\beta\gamma}$, and $T_\alpha{}^\beta{}_\gamma$ are essentially the same tensor in that they are connected by the known metric tensor or its inverse. Several indices may be raised or lowered simultaneously as the

---

† Here we have written the lower indices directly below the upper indices rather than off-set them as in (5.3). We shall oftentimes do this, provided no confusion results as in the case of raising or lowering indices.

following example shows. Let $S^{\alpha\beta}{}_{\gamma\varepsilon}$ be a tensor of type $\binom{2}{2}$. We can lower $\alpha$ and $\beta$ while raising $\varepsilon$ by taking

$$S_{\alpha\beta\gamma}{}^{\varepsilon} = S^{\eta\lambda}{}_{\gamma\mu} g_{\alpha\eta} g_{\beta\lambda} g^{\mu\varepsilon}.$$

It has been our goal in the preceding discussion to present the basic, classical definitions of tensors and some of their operations. The notation of tensors, although at first glance rather overpowering due to the many indices, is easily mastered and quite remarkable in itself. The tensor notation makes it simple, for instance, to check the validity of equations involving several indices; summed indices always occur as an upper and lower pair and free indices, or nonrepeated indices over which a sum is not taken, should occur at the same level on both sides of an equation. A more detailed account of tensors can be found in Lovelock and Rund [1] or Landau and Lifschitz [1].

## 5.3 THE LORENTZ GROUP

Einstein based his special theory of relativity upon two fundamental postulates, the Postulate of Relativity which states that all laws of nature and results of experiments performed in a given reference frame are independent of the translational motion of the system, and the Postulate of the Constancy of the Speed of Light. The second postulate permits us to deduce a connection between two space–time coordinate systems moving at constant velocity with respect to each other; this connection manifests itself in the Lorentz transformations.

The fact that the velocity of propagation of an electromagnetic disturbance in free space is a universal constant $c$ (approximately, $c = 2.998 \times 10^8$ meters per second) which is independent of the inertial frame can be exploited as follows. Suppose that a light source is fixed at the origin 0 of a coordinate system $(x, y, z)$. At time $t = 0$, if a spherical wave front is emitted from the source, then an observer at 0 will note that the wave front has an equation

$$x^2 + y^2 + z^2 - c^2 t^2 = 0.$$

On the other hand, an observer at the origin $\bar{0}$ in a second coordinate system $(\bar{x}, \bar{y}, \bar{z})$ moving with uniform velocity relative to the first observer will see, assuming 0 and $\bar{0}$ coincide at $t = \bar{t} = 0$, a spherical wave shell in his coordinate system expand with equation

$$\bar{x}^2 + \bar{y}^2 + \bar{z}^2 - c^2 \bar{t}^2 = 0,$$

since, by the second postulate, the light wave is propagated at the same speed $c$ in both systems (here, by $\bar{t}$ we mean the time measured in the barred system). The linear transformations which connect the barred to the unbarred coordinates and which preserve the quadratic form $x^2 + y^2 + z^2 - c^2t^2$ are called Lorentz transformations.

With these motivating remarks in mind, let us now be more precise in the formulation. For convenience, let us denote the space–time coordinates by

$$t^0 = ct, \qquad t^1 = x, \qquad t^2 = y, \qquad t^3 = z. \tag{5.8}$$

We can make the following definition.

**5.1   Definition**   A *general Lorentz transformation* is a linear transformation

$$\bar{t}^\alpha = \sum_{\beta=0}^{3} a_\beta{}^\alpha t^\beta \tag{5.9}$$

that leaves invariant the quadratic form

$$-(t^0)^2 + (t^1)^2 + (t^2)^2 + (t^3)^2. \tag{5.10}$$

In matrix form, the transformation may be written

$$\bar{t} = At, \tag{5.11}$$

where

$$A = (a_\beta{}^\alpha)$$

and $t = (t^0, t^1, t^2, t^3)^{\mathrm{T}}$, $\bar{t} = (\bar{t}^0, \bar{t}^1, \bar{t}^2, \bar{t}^3)^{\mathrm{T}}$, and where the superscript T denotes the operation of transpose. Note that we have abandoned our use of $t$ denoting time; here, and in the sequel, $t$ will usually denote a four-tuple of space–time coordinates. The context should make it evident.

Now we seek conditions on the matrix $A$ of the transformation under which it preserves the given quadratic form, i.e., for which

$$-(\bar{t}^0)^2 + (\bar{t}^1)^2 + (\bar{t}^2)^2 + (\bar{t}^3)^2 = -(t^0)^2 + (t^1)^2 + (t^2)^2 + (t^3)^2. \tag{5.12}$$

When (5.9) is substituted into (5.12), we obtain

$$-\left( \sum_{\beta=0}^{3} a_\beta{}^0 t^\beta \right)^2 + \sum_{\alpha=1}^{3} \left( \sum_{\beta=0}^{3} a_\beta{}^\alpha t^\beta \right)^2 = -(t^0)^2 + \sum_{\alpha=1}^{3} (t^\alpha)^2.$$

Multiplying out the left-hand side and collecting terms we get

$$\sum_{\alpha=0}^{3}\sum_{\beta=0}^{3}\left(-a_\beta{}^0 a_\alpha{}^0 + \sum_{\gamma=1}^{3} a_\alpha{}^\gamma a_\beta{}^\gamma\right)t^\alpha t^\beta = -(t^0)^2 + \sum_{\alpha=1}^{3}(t^\alpha)^2.$$

Consequently, we obtain the defining relations for a general Lorentz transformation:

$$-a_\beta{}^0 a_\alpha{}^0 + \sum_{\gamma=1}^{3} a_\alpha{}^\gamma \alpha_\beta{}^\gamma = \begin{cases} -1, & \text{if } \alpha = \beta = 0, \\ 1, & \text{if } \alpha = \beta > 0, \\ 0, & \text{if } \alpha \neq \beta. \end{cases} \tag{5.13}$$

In matrix form, these conditions may be expressed as

$$\hat{A} A = \hat{I}, \tag{5.14}$$

where

$$\hat{A} = \begin{bmatrix} -a_0{}^0 & a_0{}^1 & a_0{}^2 & a_0{}^3 \\ -a_1{}^0 & a_1{}^1 & a_1{}^2 & a_1{}^3 \\ -a_2{}^0 & a_2{}^1 & a_2{}^2 & a_2{}^3 \\ -a_3{}^0 & a_3{}^1 & a_3{}^2 & a_3{}^3 \end{bmatrix}, \qquad \hat{I} = \begin{bmatrix} -1 & & & \\ & 1 & & \\ & & 1 & \\ & & & 1 \end{bmatrix}.$$

From (5.14), it follows at once that

$$\det \hat{A} \det A = -1,$$

and thus

$$(\det A)^2 = 1,$$

from which we conclude that

$$\det A = 1 \quad \text{or} \quad \det A = -1. \tag{5.15}$$

Thus the determinant of a general Lorentz transformation has magnitude one. It is not difficult to see that the set of all general Lorentz transformations form a group under composition, i.e., when the group multiplication $AB$ of the Lorentz transformations $A$ and $B$ is defined as the transformation obtained by successive application of $B$ then $A$. This group is known as the *general Lorentz group*.

When $\alpha = \beta = 0$ in Eq. (5.13), we obtain

$$(a_0{}^0)^2 = 1 + \sum_{\gamma=1}^{3}(a_0{}^\gamma)^2 \geq 1,$$

from which we infer that

$$a_0{}^0 \geq 1 \qquad \text{or} \qquad a_0{}^0 \leq -1.$$

As a result, we make the following definition.

**5.2   Definition**   A *Lorentz transformation* is a general Lorentz transformation satisfying the condition

$$a_0{}^0 \geq 1. \tag{5.16}$$

Our motivation for this choice lies in the fact that Lorentz transformations, characterized by condition (5.16), convert positive timelike vectors into positive timelike vectors (see Exercise 5.2). Again, it is easy to see that the Lorentz transformations form a group. Noting that condition (5.15) partitions the general Lorentz transformations into two categories, we make one further definition.

**5.3   Definition**   A *proper Lorentz transformation* is a Lorentz transformation satisfying the condition

$$\det A = 1.$$

The set of all proper Lorentz transformations also forms a group, a subgroup of the general Lorentz group, called the *proper Lorentz group.*

Before considering infinitesimal representations of the proper Lorentz group, which is relevant from our point of view, we wish to make a few remarks concerning terminology. Technically, the proper Lorentz group as we have defined it perhaps should be more precisely termed the proper *homogeneous* Lorentz group. The reason for this is that we could define the nonhomogeneous transformations which leave the quadratic form

$$-(t^0 - \tau^0)^2 + (t^1 - \tau^1)^2 + (t^2 - \tau^2)^2 + (t^3 - \tau^3)^2$$

invariant; these contain the proper Lorentz transformations ("rotations") as well as space–time translations. Such transformations form the so-called *inhomogeneous Lorentz group* or the *Poincaré group* and are characterized by a nonhomogeneous term in the linear transformation due to the translation. Hereafter, when we refer to a Lorentz transformation, we mean a proper, homogeneous Lorentz transformation.

For our purposes we need to write down the proper Lorentz transformations in terms of arbitrary parameters so that our theorems on invariant variation problems can be readily applied. To this end, we note that to determine the matrix $A$ of the transformation we must determine the sixteen elements $a_\beta{}^\alpha$. These elements satisfy the defining relations (5.13) which represent sixteen equations; however, the equations (5.13) are obviously not independent since when $\alpha$ and $\beta$ are interchanged the same equation results. Consequently, there are $16 - 6 = 10$ independent equations among (5.13) for the sixteen elements of $A$. We can therefore conclude that there are six free parameters which must be specified to determine the group. In other words, once we specify six of the $a^\alpha{}_\beta$, then the remaining ten $a^\alpha{}_\beta$ are determined from the ten independent equations in (5.13). We shall now describe how these six free parameters can be chosen and what their relationship is to physical problems.

If we consider a proper Lorentz transformation $A$ which does not change the time coordinate $t^0$, then $A$ must be of the form

$$A = \begin{bmatrix} 1 & 0 & 0 & 0 \\ 0 & a_1{}^1 & a_2{}^1 & a_3{}^1 \\ 0 & a_1{}^2 & a_2{}^2 & a_3{}^2 \\ 0 & a_1{}^3 & a_2{}^3 & a_3{}^3 \end{bmatrix}, \tag{5.17}$$

and it must preserve the quadratic form

$$(t^1)^2 + (t^2)^2 + (t^3)^2,$$

which represents the ordinary Euclidean distance from the origin in three dimensional space. Evidently, the set of all such transformations (5.17) forms a subgroup of the proper Lorentz group which is isomorphic to the three-dimensional rotation group (the set of all rotations in three dimensional space about a fixed point). It is well known (see Goldstein [1]) that such rotations are determined fully by three independent parameters representing angles of rotation about the $t^1$-, $t^2$-, and $t^3$-axes, respectively. Hence, in space–time, let us consider a rotation about the $t^1$-axes; if $\varepsilon$ is the angle of rotation, then the transformation formulas will be

$$\begin{aligned} \bar{t}^0 &= t^0, \qquad \bar{t}^1 = t^1, \\ \bar{t}^2 &= t^2 \cos \varepsilon + t^3 \sin \varepsilon, \\ \bar{t}^3 &= -t^3 \sin \varepsilon + t^2 \cos \varepsilon. \end{aligned} \tag{5.18}$$

If we denote the matrix of this transformation by $A_1(\varepsilon)$, then

$$A_1(\varepsilon) = \begin{bmatrix} 1 & 0 & 0 & 0 \\ 0 & 1 & 0 & 0 \\ 0 & 0 & \cos \varepsilon & \sin \varepsilon \\ 0 & 0 & -\sin \varepsilon & \cos \varepsilon \end{bmatrix} \tag{5.19}$$

Similarly, the matrices $A_2(\varepsilon)$ and $A_3(\varepsilon)$ representing rotations about the $t^2$ and $t^3$-axis in space–time are

$$A_2(\varepsilon) = \begin{bmatrix} 1 & 0 & 0 & 0 \\ 0 & \cos \varepsilon & 0 & \sin \varepsilon \\ 0 & 0 & 1 & 0 \\ 0 & -\sin \varepsilon & 0 & \cos \varepsilon \end{bmatrix}, \qquad A_3(\varepsilon) = \begin{bmatrix} 1 & 0 & 0 & 0 \\ 0 & \cos \varepsilon & \sin \varepsilon & 0 \\ 0 & -\sin \varepsilon & \cos \varepsilon & 0 \\ 0 & 0 & 0 & 1 \end{bmatrix}, \qquad (5.20)$$

respectively.

Let us now investigate the form of a proper Lorentz transformation when the time coordinate $t^0$ is also subjected to the transformation. The simplest cases are clearly the ones in which a single spatial coordinate either $t^1$, $t^2$, or $t^3$, is also transformed, the remaining two being left unchanged. Therefore, let us consider the proper Lorentz transformation $A = (a_\beta{}^\alpha)$ which transforms $t^0$ and $t^1$ but leaves $t^2$ and $t^3$ unchanged. Clearly, $A$ must be of the form

$$A = \begin{bmatrix} a_0{}^0 & a_1{}^0 & 0 & 0 \\ a_0{}^1 & a_1{}^1 & 0 & 0 \\ 0 & 0 & 1 & 0 \\ 0 & 0 & 0 & 1 \end{bmatrix}. \qquad (5.21)$$

The defining conditions (5.13) give the relations

$$\begin{aligned} (a_0{}^1)^2 - (a_0{}^0)^2 &= -1, \\ (a_1{}^1)^2 - (a_1{}^0)^2 &= 1, \\ a_0{}^1 a_1{}^1 - a_1{}^0 a_0{}^0 &= 0. \end{aligned} \qquad (5.22)$$

From the last equation in (5.22) we obtain

$$a_0{}^0 = k a_1{}^1, \qquad a_0{}^1 = k a_1{}^0 \qquad (k \text{ constant}).$$

When these values are substituted into (5.22b) we obtain, after using (5.22a), the condition $k = \pm 1$. But $k = -1$ is ruled out since $\det A = 1$, and so

$$a_0{}^0 = a_1{}^1, \qquad a_0{}^1 = a_1{}^0.$$

Therefore, from Eqs. (5.22), we may put

$$a_0{}^0 = a_1{}^1 = \cosh \varepsilon, \qquad a_0{}^1 = a_1{}^0 = \sinh \varepsilon$$

for some parameter $\varepsilon$. The matrix of the transformation leaving $t^2$ and $t^3$ fixed is therefore

$$A_4(\varepsilon) = \begin{bmatrix} \cosh \varepsilon & \sinh \varepsilon & 0 & 0 \\ \sinh \varepsilon & \cosh \varepsilon & 0 & 0 \\ 0 & 0 & 1 & 0 \\ 0 & 0 & 0 & 1 \end{bmatrix}. \qquad (5.23)$$

Exactly similar arguments yield matrices $A_5(\varepsilon)$ and $A_6(\varepsilon)$ given by

$$A_5(\varepsilon) = \begin{bmatrix} \cosh \varepsilon & 0 & \sinh \varepsilon & 0 \\ 0 & 1 & 0 & 0 \\ \sinh \varepsilon & 0 & \cosh \varepsilon & 0 \\ 0 & 0 & 0 & 1 \end{bmatrix} \qquad (5.24)$$

and

$$A_6(\varepsilon) = \begin{bmatrix} \cosh \varepsilon & 0 & 0 & \sinh \varepsilon \\ 0 & 1 & 0 & 0 \\ 0 & 0 & 1 & 0 \\ \sinh \varepsilon & 0 & 0 & \cosh \varepsilon \end{bmatrix}, \qquad (5.25)$$

which leave $t^1$, $t^3$, and $t^1$, $t^2$ fixed, respectively.

From the foregoing it is clear that we can make the interpretation that of the six free parameters of the proper Lorentz transformation, three specify the orientation of the spatial coordinate axes and the remaining three specify the relative velocity between the barred and unbarred coordinate frames.

## 5.4    INFINITESIMAL LORENTZ TRANSFORMATIONS

Let $A(\varepsilon) = (a_{ij}(\varepsilon))$ be the matrix of a given linear transformation depending on a parameter $\varepsilon$ such that $A(0) = I$, the identity matrix. Expanding the matrix in a power series about $\varepsilon = 0$, we get

$$A(\varepsilon) = I + \varepsilon A + \cdots, \qquad (5.26)$$

where

$$A = \frac{d}{d\varepsilon} A(\varepsilon) \bigg|_{\varepsilon = 0}. \qquad (5.27)$$

The matrix $A$ is called the *infinitesimal matrix* (or transformation) corresponding to the transformation $A(\varepsilon)$. If the infinitesimal transformation $A$ is known, then the matrix $A(\varepsilon)$ can clearly be reconstructed via

$$A(\varepsilon) = \exp(\varepsilon A). \qquad (5.28)$$

In the previous section we determined six basic transformations $A_1(\varepsilon), \ldots, A_6(\varepsilon)$ for the proper Lorentz group. Using the prescription (5.26)

and (5.27) we can compute their corresponding *infinitesimal Lorentz trans-*
*formations*; we obtain

$$A_1 = \begin{bmatrix} 0 & 0 & 0 & 0 \\ 0 & 0 & 0 & 0 \\ 0 & 0 & 0 & 1 \\ 0 & 0 & -1 & 0 \end{bmatrix}, \quad A_2 = \begin{bmatrix} 0 & 0 & 0 & 0 \\ 0 & 0 & 0 & 1 \\ 0 & 0 & 0 & 0 \\ 0 & -1 & 0 & 0 \end{bmatrix},$$

$$A_3 = \begin{bmatrix} 0 & 0 & 0 & 0 \\ 0 & 0 & 1 & 0 \\ 0 & -1 & 0 & 0 \\ 0 & 0 & 0 & 0 \end{bmatrix}, \quad A_4 = \begin{bmatrix} 0 & 1 & 0 & 0 \\ 1 & 0 & 0 & 0 \\ 0 & 0 & 0 & 0 \\ 0 & 0 & 0 & 0 \end{bmatrix}, \quad (5.29)$$

$$A_5 = \begin{bmatrix} 0 & 0 & 1 & 0 \\ 0 & 0 & 0 & 0 \\ 1 & 0 & 0 & 0 \\ 0 & 0 & 0 & 0 \end{bmatrix}, \quad A_6 = \begin{bmatrix} 0 & 0 & 0 & 1 \\ 0 & 0 & 0 & 0 \\ 0 & 0 & 0 & 0 \\ 1 & 0 & 0 & 0 \end{bmatrix}.$$

Now, if $A$ is an arbitrary infinitesimal proper Lorentz transformation, then
it must be a linear combination of $A_1, \ldots, A_6,$† i.e.,

$$A = \varepsilon_{01} A_4 + \varepsilon_{02} A_5 + \varepsilon_{03} A_6 + \varepsilon_{12} A_3 + \varepsilon_{13} A_2 + \varepsilon_{23} A_1,$$

where $\varepsilon_{01}$, $\varepsilon_{02}$, $\varepsilon_{03}$, $\varepsilon_{12}$, $\varepsilon_{13}$, and $\varepsilon_{23}$ are arbitrary real parameters. Consequently, a Lorentz transformation can be written

$$\bar{t} = \exp(A)t$$
$$= (I + A + \cdots)t,$$

or, in explicit component form,

$$\begin{bmatrix} \bar{t}^0 \\ \bar{t}^1 \\ \bar{t}^2 \\ \bar{t}^3 \end{bmatrix} = \begin{bmatrix} t^0 \\ t^1 \\ t^2 \\ t^3 \end{bmatrix} + \begin{bmatrix} 0 & \varepsilon_{01} & \varepsilon_{02} & \varepsilon_{03} \\ \varepsilon_{01} & 0 & \varepsilon_{12} & \varepsilon_{13} \\ -\varepsilon_{02} & -\varepsilon_{12} & 0 & \varepsilon_{23} \\ \varepsilon_{03} & -\varepsilon_{13} & \varepsilon_{23} & 0 \end{bmatrix} \begin{bmatrix} t^0 \\ t^1 \\ t^2 \\ t^3 \end{bmatrix}. \quad (5.30)$$

† We have not explicitly shown that any infinitesimal Lorentz matrix lies in a six-dimen-
sional space. Actually there is a well-developed theory, the theory of Lie groups, from which our
results in this section would easily follow; however, our choice has been to give a strictly classical
discussion of these matters. But, briefly, the proper Lorentz group can be regarded as a six-
dimensional manifold which is a submanifold of the sixteen dimensional manifold GL(4), the
general linear group consisting of all 4 × 4 matrices. A one-parameter family of transformations
$A(\varepsilon)$ satisfying $A(0) = I$ is a curve in a group which passes through the identity. Its tangent space
is a vector space of the same dimension as the manifold. Hence, according to (5.26) and (5.27),
the infinitesimal matrices lie in the tangent space at the identity, and with appropriate commuta-
tion rules, form the Lie algebra of the Lie group. The reader interested in pursuing these ideas
should consult Spivak [1].

Here, we have neglected the higher-order terms. Still more concisely we can write (5.30) in terms of the *Minkowski metric*

$$g^{00} = -1, \qquad g^{11} = g^{22} = g^{33} = 1,$$
$$g^{\alpha\beta} = 0 \qquad (\alpha \neq \beta) \tag{5.31}$$

as

$$\bar{t}^{\alpha} = t^{\alpha} + \sum_{\beta=0}^{3} g^{\beta\beta}\varepsilon_{\alpha\beta}t^{\beta}, \tag{5.32}$$

where $\varepsilon_{\alpha\beta} = -\varepsilon_{\beta\alpha}$. Equation (5.32) is the equation that will be of importance in the subsequent discussion.

By choosing coordinates different from those selected in (5.8), the Lorentz transformation can be made to appear as an ordinary rotation of four-dimensional space. In particular, if we allow complex coordinates and choose

$$t^0 = ict, \quad t^1 = x, \quad t^2 = y, \quad t^3 = z, \tag{5.33}$$

then the invariant quadratic form becomes

$$(t^0)^2 + (t^1)^2 + (t^2)^2 + (t^3)^2.$$

It is well known (see Näimark [1], for example) that the linear transformations leaving this form invariant are the rotations, or the orthogonal group. By repeating many of the arguments above, we can eventually write down the infinitesimal Lorentz transformations as

$$\bar{t}^{\alpha} = t^{\alpha} + \varepsilon_{\alpha\beta}t^{\alpha} \qquad \text{(complex time coordinate)}, \tag{5.34}$$

where $\varepsilon_{\alpha\beta} = -\varepsilon_{\beta\alpha}$. Although this method of writing space–time coordinates is presently losing some of its popularity, it still is useful in simplifying the notation used in complicated field theories. This simplification arises, of course, from the fact that in complex coordinates the metric tensor reduces to the Kronecker delta, i.e.,

$$g_{\alpha\beta} = \begin{cases} 0, & \alpha \neq \beta, \\ 1, & \alpha = \beta. \end{cases}$$

## 5.5   PHYSICAL FIELDS

We consider a physical system which is characterized by a set of $n$ functions $x^1(t^0, \ldots, t^m), \ldots, x^n(t^0, \ldots, t^m)$ depending upon the time coordinate $t^0$ and upon $m$ spatial coordinates $t^1, \ldots, t^m$. The $n$ functions, which we write simply as $x^1(t), \ldots, x^n(t)$, where $t = (t^0, \ldots, t^m)$, will be called the *field functions* of the system. Our basic supposition is that the field functions $x^1(t), \ldots, x^n(t)$ satisfy a system of equations that are derivable as necessary conditions for a given functional to be extremal; that is to say, we assume that the $x^1(t), \ldots, x^n(t)$ are solutions to the Euler–Lagrange equations

$$\frac{\partial L}{\partial x^k} - \frac{\partial}{\partial t^\alpha} \frac{\partial L}{\partial \dot{x}_\alpha^k} = 0 \qquad (k = 1, \ldots, n) \tag{5.35}$$

corresponding to a fundamental integral

$$J(x(t)) = \int_D L(x(t), \partial x(t)) \, dt^0 \cdots dt^m, \tag{5.36}$$

where, in the notation of the preceding chapters, $x(t) = (x^1(t), \ldots, x^n(t))$ and $D$ is a given region in $R^{m+1}$. In this context, the function $L$ is called the *Lagrangian density* and the fundamental integral (5.36) is the *action integral* of the field. Note that we have implicitly assumed in (5.36) that the Lagrangian density does not depend directly on the coordinates $t^0, \ldots, t^m$. When $n = 1$, i.e., there is only one field function, we shall say that the field is a scalar field. A choice of $n > 1$ will correspond to vector fields, tensor fields, and so on, depending upon the transformation law obeyed by the $x^k(t)$. For covariant vector fields we shall write the index down, and for tensor fields we shall use multiple indices. For the most part, we shall have $m = 3$, corresponding to space–time coordinates. Finally, the Euler–Lagrange equations (5.35) will be referred to as the *field equations* for the system.

**5.1   Example**   Let $n = 1$ and $m = 3$ with $(t^0, t^1, t^2, t^3) = (t, x, y, z)$, and let the scalar field function be denoted by $u = u(t, x, y, z)$. If this scalar field describes uncharged particles of mass $m$ and spin zero, then the governing field equation is the Klein–Gordon equation

$$\frac{\partial^2 u}{\partial t^2} - \nabla^2 u = m^2 u, \tag{5.37}$$

where $\nabla^2$ is the Laplacian operator defined by

$$\nabla^2 \equiv \frac{\partial^2}{\partial x^2} + \frac{\partial^2}{\partial y^2} + \frac{\partial^2}{\partial z^2}.$$

In this case, the corresponding action integral is given by

$$J(u) = \int_D \tfrac{1}{2}(u_t^2 - u_x^2 - u_y^2 - u_z^2 - m^2 u^2) \, dt \, dx \, dy \, dz, \tag{5.38}$$

where $D$ is a region in space–time and $u_t = \partial u/\partial t, \ldots$, etc. It is easily checked that Eq. (5.37) is the Euler–Lagrange equation corresponding to the functional (5.38).

**5.2   Example**   Let $n = 4$ and $m = 3$ with $(t^0, t^1, t^2, t^3) = (t, x, y, z)$ and denote the four field functions by $A_1(t, x, y, z), \ldots, A_4(t, x, y, z)$. If these field functions represent a four-potential for the electromagnetic field, then the governing equations, Maxwell's equations, take the form

$$\frac{\partial^2 A_k}{\partial t^2} - \nabla^2 A_k = 0 \qquad (k = 1, 2, 3, 4),$$

with the action integral given by

$$J(A_1, A_2, A_3, A_4) = \int_D \left[ \frac{1}{2} \left( \text{grad } A_1 - \frac{\partial \mathbf{A}}{\partial t} \right)^2 - \frac{1}{2} (\text{curl } \mathbf{A})^2 \right] dt \, dx \, dy \, dz,$$

where $A_1$ is the scalar potential and $\mathbf{A} = (A_2, A_3, A_4)$ is the vector potential. This example will be developed in detail in a subsequent section.

We are now able to make some general remarks regarding the invariance requirements which must be imposed upon variational or action principles for physical fields. The basis for such remarks lies in the fact that any field theory must be consistent with the special theory of relativity; in view of the discussions presented in the preceding sections we can write down some basic assumptions or conditions which must be obeyed by the action integral for a given physical field:

*Assumption 1*   The field functions $x^1(t), \ldots, x^n(t)$ must satisfy the Euler–Lagrange equations (5.35) corresponding to the action integral (5.36).

*Assumption 2*   The action integral for a physical field must be absolutely invariant under the ten-parameter Poincaré group consisting of space–time translations

$$\bar{t}^\alpha = t^\alpha + \varepsilon^\alpha \tag{5.39}$$

and the proper Lorentz group

$$\bar{t}^\alpha = t^\alpha + g^{\beta\beta} \varepsilon_{\alpha\beta} t^\beta, \tag{5.40}$$

where $g^{\alpha\beta}$ is the Minkowski metric given by (5.31). The ten parameters are $\varepsilon^0, \ldots, \varepsilon^3, \varepsilon_{01}, \varepsilon_{02}, \varepsilon_{03}, \varepsilon_{12}, \varepsilon_{13}$, and $\varepsilon_{23}$. Assumption 1 provides a variational principle for the field; it essentially states that $\delta \int_D L \, dt^0 \, dt^1 \, dt^2 \, dt^3 = 0$. Assumption 2 expresses the invariance property under Lorentz transformations and space–time translations. (Actually, our assumption of invariance under translations is redundant since we have explicitly assumed that the Lagrangian density $L$ does not depend directly upon the $t^\alpha$ coordinates.)

**5.1   Remark**   It is easily seen either in the case of (5.39) or (5.40) that the determinant

$$\Delta = \det\left( \frac{\partial \bar{t}^\alpha}{\partial t^\beta} \right) = 1. \tag{5.41}$$

Consequently, using Definition 4.1, it follows that the invariance of the action integral under the Poincaré group is equivalent to invariance of the Lagrangian density.

Assumption 2 also provides a ready-made ten-parameter family of transformations under which we can study invariant variation problems for physical fields. If the Lagrangian meets the test of invariance, and this must be so for any consistent theory, then Noether's theorem can be applied immediately to directly determine the conservation laws for the field. This method has proved to be a powerful tool in the study of modern field theories, and it also points out an advantage of a variational formulation of field theory. In addition to providing instant access to conserved quantities, the invariance hypothesis expressed in Assumption 2 also gives a method for determining or characterizing possible Lagrangians for a given field; not every Lagrangian is possible, but only those satisfying the invariance condition

$$\bar{L} - L = o(\varepsilon).$$

Therefore, by the remarks in the last chapter, the Lagrangian $L$ must satisfy the fundamental invariance identities (4.14).

We shall end this section with a derivation of concise tensor formulas for the *infinitesimal generators* of the Poincaré group given by (5.39) and (5.40). It is clear from (5.39) that in the case of translations we have

$$\tau_\beta{}^\alpha = \frac{\partial \bar{t}^\alpha}{\partial \varepsilon^\beta}\bigg|_0 = \delta_\beta{}^\alpha. \tag{5.42}$$

For the proper homogeneous Lorentz transformations (5.40), we can see that the generators are

$$\tau^\alpha_{\lambda\mu} \equiv \frac{\partial \bar{t}^\alpha}{\partial \varepsilon_{\lambda\mu}}\bigg|_0 \qquad (\lambda < \mu),$$

$$= g^{\mu\mu}\delta_{\alpha\lambda}t^\mu - g^{\lambda\lambda}\delta_{\alpha\mu}t^\lambda \qquad \text{(no sums)}, \tag{5.43}$$

where $(\lambda, \mu) = (0, 1), (0, 2), (0, 3), (1, 2), (1, 3),$ and $(2, 3)$. Hereafter, we shall denote these six ordered pairs by the set symbol $S$.

Before embarking upon a detailed study of scalar fields, we wish to make one final, but important, remark concerning the invariance of the field functions. It is clear that the group generators $\tau_\beta{}^\alpha$ and $\tau^\alpha_{\lambda\mu}$ are derivable from the defining transformations of the Lorentz group. On the other hand, the generators $\xi_\beta{}^\alpha$ and $\xi^\alpha_{\lambda\mu}$ of the transformations $x^\alpha \to \bar{x}^\alpha$ will depend upon the transformation law of the field functions $x^1(t), \ldots, x^n(t)$ under the Lorentz group which, in turn, is determined solely by the tensorial character of the field functions. This comment will become relevant when we examine special fields (e.g., scalar fields and covariant vector fields).

## 5.6   SCALAR FIELDS

A scalar field is described by a single field function $x(t) = x(t^0, t^1, t^2, t^3)$ which is invariant (a scalar) under the transformations (5.39) and (5.40), i.e., $\bar{x}(\bar{t}) = x(t)$. The action integral for such a field is of the form

$$J = \int_D L(x(t), \dot{x}_0(t), \dot{x}_1(t), \dot{x}_2(t), \dot{x}_3(t))\, dt^0 \cdots dt^3, \qquad (5.44)$$

where $\dot{x}_\alpha(t) \equiv \partial x(t)/\partial t^\alpha$, $\alpha = 0, \ldots, 3$, and it is by our assumptions invariant under the transformations

$$\bar{t}^\alpha = t^\alpha + \varepsilon^\alpha, \qquad \bar{x} = x \qquad (5.45)$$

and under the transformations

$$\bar{t}^\alpha = t^\alpha + g^{\beta\beta}\varepsilon_{\alpha\beta} t^\beta, \qquad \bar{x} = x, \qquad (5.46)$$

where, as before, $\alpha, \beta, \ldots$ range over 0 through 3. We note that the transformation $x \to \bar{x} = x$ is the identity transformation since the field is scalar and therefore does not change when the space–time coordinates are transformed.

According to Noether's theorem, it follows that conservation laws for scalar fields may be written down directly. These are summarized in the following theorem.

**5.1   Theorem**   For a scalar field $x(t)$, the following ten conservation laws hold true:

(i)   $\dfrac{\partial}{\partial t^\alpha}(\dot{x}_\beta L_{\dot{x}_\alpha} - L\delta_\beta{}^\alpha) = 0$        $(\beta = 0, \ldots, 3)$,        $(5.47)$

(ii)   $\dfrac{\partial}{\partial t^\alpha}(H_\lambda{}^\alpha g^{\mu\mu}t^\mu - H_\mu{}^\alpha g^{\lambda\lambda}t^\lambda) = 0$        $(\lambda, \mu) \in S$,        $(5.48)$

where

$$H_\beta{}^\alpha \equiv -L\delta_\beta{}^\alpha + \dot{x}_\beta L_{\dot{x}_\alpha} \qquad (5.49)$$

and no summation is intended over $\lambda$ and $\mu$ in (5.48).

The subscript $\dot{x}_\alpha$ on $L$ denotes partial differentiation with respect to $\dot{x}_\alpha$. To prove (5.47) we notice that the generators of the transformation (5.45) are

$$\tau_\beta{}^\alpha \equiv \frac{\partial \bar{t}^\alpha}{\partial \varepsilon^\beta}\bigg|_0 = \delta_\beta{}^\alpha$$

and

$$\xi_\beta \equiv \frac{\partial \bar{x}}{\partial \varepsilon^\beta}\bigg|_0 = 0.$$

Substitution of these quantities in (4.14) gives (5.47) at once. To check the validity of (5.48), we first note that the generators of the transformation (5.46) are given by

$$\tau^\alpha_{\lambda\mu} = g^{\mu\mu}\delta_{\alpha\lambda}t^\mu - g^{\lambda\lambda}\delta_{\alpha\mu}t^\lambda \qquad \text{(no sums)}$$

and

$$\xi_{\lambda\mu} \equiv \frac{\partial \bar{x}}{\partial \varepsilon_{\lambda\mu}}\bigg|_0 = 0.$$

Substitution into (4.14) gives

$$\frac{\partial}{\partial t^\alpha}\left((-L\delta_\beta{}^\alpha + \dot{x}_\beta L_{\dot{x}_\alpha})\tau^\beta_{\lambda\mu}\right) = 0,$$

which immediately reduces to (5.48), completing the proof of the theorem.

□

The quantities $H_\beta{}^\alpha$ defined by (5.49) form the components of the so-called *energy–momentum tensor* for the scalar field. In terms of this tensor, the conservation law (5.47) may be written

$$\frac{\partial}{\partial t^\alpha} H_\beta{}^\alpha = 0.$$

Recalling expressions (4.33), which define the conserved quantities, or field invariants, of the field, we conclude from (5.47) that the four quantities

$$\Pi_\beta \equiv \iiint H_\beta{}^0 \, dt^1 \, dt^2 \, dt^3 \tag{5.50}$$

are independent of time and hence represent conserved quantities for the scalar field. $H_0{}^0$ is called the energy density and $H_1{}^0, H_2{}^0, H_3{}^0$ are called the momentum densities of the field.

In a similar manner, let us define the *angular momentum tensor* by

$$M^\alpha_{\lambda\mu} \equiv H_\lambda{}^\alpha g^{\mu\mu}t^\mu - H_\mu{}^\alpha g^{\lambda\lambda}t^\lambda \tag{5.51}$$

for $\alpha = 0, \ldots, 3$ and $(\lambda, \mu) \in S$. It then follows from (5.48) and (4.33) that the six quantities

$$M_{\lambda\mu} \equiv \iiint M_{\mu\lambda}^0 \, dt^1 \, dt^2 \, dt^3, \qquad (\lambda, \mu) \in S, \tag{5.52}$$

are field invariants.

Equations (5.50) and (5.52) are summarized by saying that the energy $\Pi_0$, momentum $\Pi_1$, $\Pi_2$, $\Pi_3$, and angular momentum $M_{\lambda\mu}$ of the field are conserved. The degree to which these interpretations are valid is illustrated by focusing our attention to a previous example.

**5.3   Example**   From Example 5.1, the Lagrangian density which leads to the Klein–Gordon equation can be written as

$$L = -\tfrac{1}{2} \sum_{\alpha=0}^{3} g^{\alpha\alpha} \dot{x}_\alpha(t)^2 - \tfrac{1}{2} m^2 x(t)^2. \tag{5.53}$$

Then, the energy–momentum tensor is given by

$$H_\beta{}^\alpha = -L\delta_\beta{}^\alpha - \dot{x}_\beta \dot{x}_\alpha g^{\alpha\alpha} \qquad \text{(no sums).}$$

Consequently,

$$H_\beta{}^0 = -L\delta_\beta{}^0 + \dot{x}_\beta \dot{x}_0 \tag{5.54}$$

and, from (5.53) and (5.54),

$$H_0{}^0 = \tfrac{1}{2} \sum_{\alpha=0}^{3} \dot{x}_\alpha(t)^2 + \tfrac{1}{2} m^2 x(t)^2,$$

which is obviously an energy-like quantity. Also from (5.54),

$$H_1{}^0 = \dot{x}_0 \dot{x}_1, \qquad H_2{}^0 = \dot{x}_0 \dot{x}_2, \qquad H_3{}^0 = \dot{x}_0 \dot{x}_3$$

give the momentum densities. We leave as an exercise the derivation of the angular momentum densities.

The next question we can ask is: To what extent does being Lorentz invariant characterize the Lagrangian of a scalar field? The fundamental invariance identities for multiple integral problems gives a partial answer. For a scalar field the generators $\xi_s{}^k$ of the transformation $x^k \rightarrow \bar{x}^k = x^k$ vanish identically so that the fundamental invariance identities (4.14) become

$$\frac{\partial L}{\partial t^\alpha} \tau_s{}^\alpha - \frac{\partial L}{\partial \dot{x}_\alpha} \frac{d\tau_s{}^\beta}{dt^\alpha} \dot{x}_\beta + L \frac{d\tau_s{}^\alpha}{dt^\alpha} = 0, \tag{5.55}$$

where the transformations are of the form

$$\bar{t}^\alpha = t^\alpha + \tau_s{}^\alpha \varepsilon^s, \qquad \bar{x}^k = x^k.$$

In the case of the translations the generators are given by (5.42) so that the invariance identity (5.55) becomes

$$\frac{\partial L}{\partial t^\alpha} = 0. \tag{5.56}$$

That is to say, the Lagrangian does not depend explicitly on the space–time coordinates. Therefore, our assumption in the last section that the Lagrangian for a physical field should not depend upon the $t^\alpha$ is certainly justified and is, in fact, redundant. For the homogeneous Lorentz transformations (5.40), the generators are given by (5.43) and we have

$$\frac{d\tau^\alpha_{\lambda\mu}}{dt^\beta} = g^{\mu\mu}\delta_{\alpha\lambda}\delta_{\mu\beta} - g^{\lambda\lambda}\delta_{\alpha\mu}\delta_{\lambda\beta} \qquad \text{(no sums)} \tag{5.57}$$

and

$$\frac{d\tau^\alpha_{\lambda\mu}}{dt^\alpha} = 0 \tag{5.58}$$

for $(\lambda, \mu) \in S$. Substitution of (5.57) and (5.58) into (5.55) gives

$$g^{\mu\mu}\frac{\partial L}{\partial \dot{x}_\mu}\dot{x}_\lambda - g^{\lambda\lambda}\frac{\partial L}{\partial \dot{x}_\lambda}x_{\dot{\mu}} = 0 \qquad \text{(no sums)} \tag{5.59}$$

for $(\lambda, \mu) \in S$. We can summarize our results in the following theorem which gives us a test for determining invariance of scalar fields.

**5.2  Theorem**  If the action integral (5.44) is Lorentz invariant, then the Lagrangian must satisfy (5.59).  □

It is easily checked, for example, that the Lagrangian density defined by (5.53) satisfies (5.59).

## 5.7  THE ELECTROMAGNETIC FIELD

In this section we shall outline some of the principal features of one of the most fundamental and aesthetic theories in nature, that of electro-magnetism. Several outstanding introductions and treatises exist on this subject; among them we mention Jackson [1] and Landau and Lifschitz [1], both of which can supply the reader with further details in the same spirit as the present development. The theory of the electromagnetic field, which arose out of the research of Faraday, Maxwell, Hemholtz, and others, provides us with an example of a covariant vector field in the setting of Minkowski space (four-dimensional space–time with the Minkowski metric). Our goal in the following paragraphs is to formulate a variational principle for electromagnetism in free space, i.e., in a vacuum with no charges present. Once this formulation is complete, then Noether's theorem can be used to derive the conservation laws for the field in much the same manner as was done for scalar fields in the last section.

In free space, the electromagnetic field can be described by two vectors:

$$\mathbf{E}(t^0, t^1, t^2, t^3) = (E_1(t^0, t^1, t^2, t^3), E_2(t^0, t^1, t^2, t^3), E_3(t^0, t^1, t^2, t^3))$$

and

$$\mathbf{B}(t^0, t^1, t^2, t^3) = (B_1(t^0, t^1, t^2, t^3), B_2(t^0, t^1, t^2, t^3), B_3(t^0, t^1, t^2, t^3)),$$

where $t^0 = ct$, and where $(t^1, t^2, t^3) = (x, y, z)$. The $\mathbf{E}$ field is known as the electric field intensity and is the electric force per unit charge; the $\mathbf{B}$ field is the magnetic force per unit charge.† The total electromagnetic force on a charge $q$ moving with velocity $\mathbf{v}$ is given by the Lorentz force law $\mathbf{F} = q(\mathbf{E} + \mathbf{v} \times \mathbf{B})$. We assume that the fields are source-free, i.e.,

$$\text{div } \mathbf{E} = 0, \qquad \text{div } \mathbf{B} = 0 \tag{5.60}$$

where div is the three-dimensional divergence operator.

In order to give a heuristic motivation for a variation principle for electromagnetic fields, we shall follow an energy approach in our discussion. Such a direction will also pay off when interpreting the conservation laws that we will obtain from invariance conditions. To this end, let us assume that $\mathbf{E}$ and $\mathbf{B}$ are defined and twice continuously differentiable in some region $\Sigma$ of space and for all times $t$ in some interval $I$. Further, let $d\tau = dt^1 \, dt^2 \, dt^3$ denote a volume element in $\Sigma$. Then, in $d\tau$, the electric and magnetic energies are given by

$$du_{el} = \tfrac{1}{2}\mathbf{E} \cdot \mathbf{E} \, d\tau$$

and

$$du_{mg} = \tfrac{1}{2}\mathbf{B} \cdot \mathbf{B} \, d\tau,$$

respectively. Therefore, the total energy in $\Sigma$ is given by the integral

$$\text{Total energy in } \Sigma = \int_{\Sigma} (\tfrac{1}{2}\mathbf{E}^2 + \tfrac{1}{2}\mathbf{B}^2) \, d\tau, \tag{5.61}$$

where both terms of the integrand are energy densities, i.e., they have units energy/volume. Now, through a surface element $d\sigma$ of the boundary of $\Sigma$, the energy flow in time $dt^0$ is given by

$$(\mathbf{E} \times \mathbf{B}) \cdot \mathbf{n} \, dt^0 \, d\sigma, \tag{5.62}$$

---

† More precisely, $\mathbf{B}$ is the magnetic deflection per unit moving charge. The formulas in this chapter are expressed in Heaviside–Lorentz units (see Jackson [1]).

where **n** the outer unit normal. If energy is to be conserved, then the decrease in energy in time in the region $\Sigma$ must equal the total energy flowing out of $\Sigma$; therefore, from (5.61) and (5.62) we obtain

$$-\frac{d}{dt^0}\int_{\Sigma}(\tfrac{1}{2}\mathbf{E}^2 + \tfrac{1}{2}\mathbf{B}^2)\,d\tau = \int_{\text{Bd }\Sigma}(\mathbf{E}\times\mathbf{B})\cdot\mathbf{n}\,d\sigma$$

$$= \int_{\Sigma}\text{div}(\mathbf{E}\times\mathbf{B})\,d\tau, \qquad (5.63)$$

where we have used the Divergence Theorem of Chapter 4. The vector $\mathbf{S} = \mathbf{E}\times\mathbf{B}$ is called *Poynting's vector* and it has units of power/area so that it is an energy flux density; to repeat what we have stated above, the normal component of **S** integrated over a closed surface gives the rate at which energy is flowing out of the surface. A common interpretation of **S**, though not entirely valid, is that it points in the direction of energy flow. If energy is conserved, then the left-hand side of (5.63) is zero so that the conservation law may be expressed in the form

$$\int_{\text{Bd }\Sigma}\mathbf{S}\cdot\mathbf{n}\,d\sigma = 0$$

or

$$\text{div }\mathbf{S} = 0. \qquad (5.64)$$

Now, we recall the vector identity $\text{div}(\mathbf{E}\times\mathbf{B}) = \mathbf{B}\cdot\text{curl }\mathbf{E} - \mathbf{E}\cdot\text{curl }\mathbf{B}$. Upon substituting this quantity into (5.63) we obtain, after rearranging,

$$-\frac{d}{dt^0}\int_{\Sigma}(\tfrac{1}{2}\mathbf{E}^2 + \tfrac{1}{2}\mathbf{B}^2)\,d\tau + \int_{\Sigma}(\mathbf{E}\cdot\text{curl }\mathbf{B} - \mathbf{B}\cdot\text{curl }\mathbf{E})\,d\tau = 0.$$

Then, differentiating and collecting terms, we get

$$\int_{\Sigma}\left[\left(-\frac{\partial\mathbf{E}}{\partial t^0} + \text{curl }\mathbf{B}\right)\cdot\mathbf{E} + \left(-\frac{\partial\mathbf{B}}{\partial t^0} - \text{curl }\mathbf{E}\right)\cdot\mathbf{B}\right]d\tau = 0. \quad (5.65)$$

For (5.65) to hold it is sufficient that

$$\text{curl }\mathbf{E} + \frac{\partial\mathbf{B}}{\partial t^0} = 0, \qquad \text{curl }\mathbf{B} - \frac{\partial\mathbf{E}}{\partial t^0} = 0. \qquad (5.66)$$

These two equations, along with the two divergence equations (5.60), form a system of eight first-order partial differential equations in the unknown components $E_1, E_2, E_3, B_1, B_2, B_3$. They are called *Maxwell's equations* and they are the governing equations for the electromagnetic field. At first it may appear that this system is overdetermined since there are eight equations and only six unknowns. However, the following calculation shows that if the fields initially exist and satisfy (5.60), then equations (5.60) may be derived from (5.66) and are therefore redundant. Hence, suppose that

$$\mathbf{E}(0, t^1, t^2, t^3) = \mathbf{E}_0(t^1, t^2, t^3), \quad \mathbf{B}(0, t^1, t^2, t^3) = \mathbf{B}_0(t^1, t^2, t^3)$$

are given initial conditions with

$$\text{div } \mathbf{E}_0 = \text{div } \mathbf{B}_0 = 0.$$

Taking the divergence of Eq. (5.66i) and noting that the divergence of a curl is identically zero, we obtain $(\partial/\partial t^0) \text{ div } \mathbf{B} = 0$, or the divergence of $\mathbf{B}$ is independent of time. Since, initially, $\text{div } \mathbf{B} = \text{div } \mathbf{B}_0 = 0$, we must have $\text{div } \mathbf{B} = 0$ for all times. Similarly one can show that $\text{div } \mathbf{E} = 0$ is a consequence of (5.66ii). Consequently, we can think of Maxwell's equations, the governing electrodynamical equations, as a set of six equations consisting of (5.66). This remark will be relevant when we formulate a variational principle for the electromagnetic field.

We now attempt to write down a Lagrangian for the electromagnetic field. Our requirements are (a) the Lagrangian should be a scalar since the fundamental integral must be scalar valued, (b) the corresponding Euler–Lagrange equations should coincide with the governing electrodynamical equations, Maxwell's equations, and (c) and action integral should be absolutely invariant under Lorentz transformations. Guided by a classical mechanical analog we might consider an energy integral of the form

$$J = \int_D (\tfrac{1}{2}\mathbf{E}^2 - \tfrac{1}{2}\mathbf{B}^2) \, dt^0 \, d\tau, \tag{5.67}$$

where the integral is taken over a "cylinder" $D = I \times \Sigma$ of space–time. This integral certainly satisfies requirement (a) but it is not clear that Maxwell's equations can be obtained from (5.67) since the Lagrangian, the integrand, does not contain derivatives of the components of the field functions. However, as it turns out, Eq. (5.67) does furnish an adequate action integral provided that we recast our formulation of electrodynamics in terms of an alternate form involving the so-called four-potential for the field. We shall presently carry out this formulation.

Let $\mathbf{A}(t^0, t^1, t^2, t^3) = (A_1(t^0, t^1, t^2, t^3), \ A_2(t^0, t^1, t^2, t^3), \ A_3(t^0, t^1, t^2, t^3))$ be a vector field defined in such a way that

$$\text{curl } \mathbf{A} = \mathbf{B}. \tag{5.68}$$

Substituting this expression into Eq. (5.66i) we obtain

$$\text{curl}\left( \mathbf{E} + \frac{\partial \mathbf{A}}{\partial t^0} \right) = 0.$$

From this equation we can conclude that there exists a scalar-valued potential function $A_0(t^0, t^1, t^2, t^3)$ for which

$$\mathbf{E} + \frac{\partial \mathbf{A}}{\partial t^0} = \text{grad } A_0. \tag{5.69}$$

The electromagnetic potentials, the vector potential $\mathbf{A}$ and the scalar potential $A_0$, are not uniquely determined by Eqs. (5.68) and (5.69); it is clear that the $A_\alpha$ ($\alpha = 0, 1, 2, 3$) may be replaced by any other four-potential $\bar{A}_\alpha$ given by

$$\bar{A}_\alpha = A_\alpha + \frac{\partial \phi}{\partial t^\alpha}, \tag{5.70}$$

where $\phi = \phi(t^0, t^1, t^2, t^3)$ is an arbitrary function of class $C^1$. A transformation of the type (5.70) is called a *gauge transformation* and it does not affect the electromagnetic field, i.e., the definition of the $A_\alpha$'s, and therefore the observables $\mathbf{B}$ and $\mathbf{E}$.

In terms of the four-potential $(A_0, A_1, A_2, A_3)$ we can rewrite the governing equations of the field. In this regard, let us eliminate $\mathbf{E}$ and $\mathbf{B}$ from (5.66i) using (5.68) and (5.69); upon substitution we obtain

$$\text{curl}(\text{curl } \mathbf{A}) + \left( \frac{\partial}{\partial t^0} \text{grad } A_0 - \frac{\partial^2 \mathbf{A}}{\partial t^{0\,2}} \right) = 0.$$

From the vector identity

$$\text{curl}(\text{curl } \mathbf{A}) = \text{grad div } \mathbf{A} - \nabla^2 \mathbf{A}$$

where $\nabla^2$ is the Laplacian operator

$$\nabla^2 = \frac{\partial^2}{\partial t^{1\,2}} + \frac{\partial^2}{\partial t^{2\,2}} + \frac{\partial^2}{\partial t^{3\,2}},$$

the last equation becomes

$$\frac{\partial^2 \mathbf{A}}{\partial t^{0\,2}} - \nabla^2 \mathbf{A} = -\text{grad}\left( \frac{\partial A_0}{\partial t^0} - \text{div } \mathbf{A} \right). \tag{5.71}$$

If we impose the additional *Lorentz condition*

$$\frac{\partial A_0}{\partial t^0} - \text{div } \mathbf{A} = 0, \tag{5.72}$$

then the vector Eq. (5.71) becomes

$$\frac{\partial^2 \mathbf{A}}{\partial t^{0\,2}} - \nabla^2 \mathbf{A} = 0. \tag{5.73}$$

In a similar fashion, it is easily shown that

$$\frac{\partial^2 A_0}{\partial t^{0\,2}} - \nabla^2 A_0 = 0. \tag{5.74}$$

Therefore, in terms of the four-potential $(A_0, A_1, A_2, A_3)$, the governing field equations are the wave Eqs. (5.73) and (5.74), assuming of course that the four-potential satisfies the Lorentz condition (5.72).

Therefore, we have an alternate formulation of the electromagnetic field in terms of the four-potential. If we rewrite the proposed variational integral (5.67) in terms of the $A_\alpha$, we obtain using (5.68) and (5.69),

$$J = \int_D \left[ \frac{1}{2} \left( \text{grad } A_0 - \frac{\partial \mathbf{A}}{\partial t^0} \right)^2 - \frac{1}{2} (\text{curl } \mathbf{A})^2 \right] dt^0 \, d\tau. \tag{5.75}$$

It is a straightforward calculation to show that the Euler–Lagrange equations

$$\frac{\partial L}{\partial A_\alpha} - \frac{\partial}{\partial t^\beta} \frac{\partial L}{\partial (\partial A_\alpha / \partial t^\beta)} = 0 \qquad (\alpha = 0, 1, 2, 3)$$

coincide with Eqs. (5.73) and (5.74) under the assumption that (5.72) holds true.

Before temporarily leaving the subject of variation principles for electromagnetic fields, we wish to offer one additional formulation which is convenient in many theoretical investigations. Therefore, we define the electromagnetic field tensor $F_{\alpha\beta}$ by

$$F_{\alpha\beta} \equiv \frac{\partial A_\beta}{\partial t^\alpha} - \frac{\partial A_\alpha}{\partial t^\beta} \qquad (\alpha, \beta = 0, 1, 2, 3), \tag{5.76}$$

where the $A_\alpha$ are the components of the four-potential. In terms of the components of the $\mathbf{E}$ and $\mathbf{B}$ fields, the components of $F_{\alpha\beta}$ are given by

$$(F_{\alpha\beta}) = \begin{bmatrix} 0 & -E_1 & -E_2 & -E_3 \\ E_1 & 0 & B_3 & -B_2 \\ E_2 & -B_3 & 0 & B_1 \\ E_3 & B_2 & -B_1 & 0 \end{bmatrix}. \tag{5.77}$$

The entries in this matrix were calculated directly from (5.76) using (5.68) and (5.69). If we define the metric tensor $g_{\alpha\beta}$ by the Minkowski metric

$$(g_{\alpha\beta}) = \begin{bmatrix} -1 & 0 & 0 & 0 \\ 0 & 1 & 0 & 0 \\ 0 & 0 & 1 & 0 \\ 0 & 0 & 0 & 1 \end{bmatrix}, \tag{5.78}$$

then $F^{\alpha\beta} = g^{\alpha\gamma}g^{\beta\delta}F_{\gamma\delta}$ and

$$
(F^{\alpha\beta}) = \begin{bmatrix} 0 & E_1 & E_2 & E_3 \\ -E_1 & 0 & B_3 & -B_2 \\ -E_2 & -B_3 & 0 & B_1 \\ -E_3 & B_2 & -B_1 & 0 \end{bmatrix}. \tag{5.79}
$$

One advantage of this formulation is that the governing field equations can be written concisely. It is a straightforward calculation to show that (5.66i) and (5.60ii) are equivalent to

$$
\frac{\partial F_{\alpha\beta}}{\partial t^{\gamma}} + \frac{\partial F_{\beta\gamma}}{\partial t^{\alpha}} + \frac{\partial F_{\gamma\alpha}}{\partial t^{\beta}} = 0 \qquad (\alpha, \beta, \gamma \text{ distinct}).
$$

The other pair of Maxwell's equations, Eqs. (5.66ii) and (5.60i), can be written

$$
\frac{\partial F^{\alpha\beta}}{\partial t^{\alpha}} = 0 \qquad (\beta = 0, 1, 2, 3).
$$

We also observe that

$$
F_{\alpha\beta} F^{\alpha\beta} = -2(E_1{}^2 + E_2{}^2 + E_3{}^2) + 2(B_1{}^2 + B_2{}^2 + B_3{}^2).
$$

Comparing with (5.67), we conclude that the action integral for the electromagnetic field in a vacuum can be written in terms of the electromagnetic field tensor as

$$
J = -\int_D \tfrac{1}{4} F_{\alpha\beta} F^{\alpha\beta} \, dt^0 \, d\tau. \tag{5.80}
$$

Now, in terms of the components of the four-potential $A_0, A_1, A_2$, and $A_3$, it is easily seen that (5.80) is equivalent to

$$
J = -\int_D \tfrac{1}{4} g^{\alpha\gamma} g^{\beta\delta} \left( \frac{\partial A_\beta}{\partial t^\alpha} - \frac{\partial A_\alpha}{\partial t^\beta} \right) \left( \frac{\partial A_\gamma}{\partial t^\delta} - \frac{\partial A_\delta}{\partial t^\gamma} \right) dt^0 \, d\tau. \tag{5.81}
$$

In summary, we have obtained a Lagrangian density for the electromagnetic field *in vacuo* in three different forms:

$$
L = \tfrac{1}{2}\mathbf{E}^2 - \tfrac{1}{2}\mathbf{B}^2 \tag{5.82}
$$

$$
= -\tfrac{1}{4} F_{\alpha\beta} F^{\alpha\beta} \tag{5.83}
$$

$$
= -\tfrac{1}{4} g^{\alpha\gamma} g^{\beta\delta} \left( \frac{\partial A_\beta}{\partial t^\alpha} - \frac{\partial A_\alpha}{\partial t^\beta} \right) \left( \frac{\partial A_\gamma}{\partial t^\delta} - \frac{\partial A_\delta}{\partial t^\gamma} \right). \tag{5.84}
$$

We shall return to this subject in Section 5.8 when we address the topic of conservation laws.

## 5.8   COVARIANT VECTOR FIELDS

What we have in mind at this point is to lay the general groundwork for a study of conservation laws in electromagnetic theory. As before, let $t = (t^0, \ldots, t^3)$ denote space–time coordinates, and let $x_0(t), \ldots, x_3(t)$ denote four field functions of class $C^2(D)$, where $D$ is a region in space–time. Here, we have anticipated the fact that we shall investigate covariant fields by our use of lower indices on the field functions. The action integral for the field then takes the form

$$J = \int_D L(x(t), \partial x(t))\, dt^0 \cdots dt^3, \tag{5.85}$$

where $x(t) = (x_0(t), \ldots, x_3(t))$ and $\partial x(t)$ denotes the collection of first partial derivatives $\partial x_\alpha / \partial t^\beta$.

As we have remarked, we shall require that (5.85) be invariant under transformations of the general type

$$\bar{t}^\alpha = t^\alpha + \tau_\beta{}^\alpha \varepsilon^\beta, \qquad \bar{x}_\alpha = x_\alpha + \xi_{\alpha\beta}\varepsilon^\beta, \tag{5.86}$$

where $\tau_\beta{}^\alpha = \tau_\beta{}^\alpha(t, x)$ and $\xi_{\alpha\beta} = \xi_{\alpha\beta}(t, x)$ are the generators. The transformation of the space–time coordinates $t^\alpha$ is given by the Poincaré group with generators given by (5.42) and (5.43); on the other hand, the generators $\xi_{\alpha\beta}$ depend on the transformation law of the vector field $x(t) = (x_0(t), \ldots, x_3(t))$. The $\xi_{\alpha\beta}$ can be calculated once this transformation law is known. Therefore, we make the assumption that the field functions $x_0(t), \ldots, x_3(t)$ are components of a covariant vector field; this implies

$$x_\alpha(t) = \frac{\partial \bar{t}^\gamma}{\partial t^\alpha} \bar{x}_\gamma(\bar{t}), \tag{5.87}$$

which is the transformation law for the components of a covariant vector field. An example of such a field is the four-potential of the electromagnetic field introduced in Section 5.7. We now prove a basic result.

**5.1   Lemma**   If

$$\bar{t}^\alpha = t^\alpha + \tau_\beta{}^\alpha \varepsilon^\beta$$

represents a transformation of space–time, and if $x_0(t), \ldots, x_3(t)$ are components of a covariant vector field, then the generators $\xi_{\alpha\beta}$ of the transformation

$$\bar{x}_\alpha = \bar{x}_\alpha(t, x, \varepsilon)$$

are given by

$$\zeta_{\alpha\beta} = -\frac{\partial \tau_\beta^{\ \gamma}}{\partial t^\alpha} x_\gamma. \tag{5.88}$$

*Proof*   Differentiating (5.87) with respect to $\varepsilon^\beta$ gives

$$0 = \frac{\partial^2 \bar{t}^\gamma}{\partial \varepsilon^\beta \, \partial t^\alpha} x_\gamma + \frac{\partial \bar{t}^\gamma}{\partial t^\alpha} \frac{\partial \bar{x}_\gamma}{\partial \varepsilon^\beta}.$$

Setting $\varepsilon^\alpha = 0$ for all $\alpha$, we obtain

$$0 = \frac{\partial \tau^\gamma}{\partial t^\alpha} x_\gamma + \delta_\alpha^{\ \gamma} \zeta_{\gamma\beta},$$

where we have used (4.3) and the fact that the generators $\zeta_{\gamma\beta}$ are given by $(\partial \bar{x}_\gamma / \partial \varepsilon^\beta)_{\varepsilon = 0}$. Transposing in the above equation yields (5.88) and therefore completes the proof.   □

In the case of space–time translations, it is easily seen that

$$\zeta_{\alpha\beta} = 0 \qquad (\alpha, \beta = 0, \ldots, 3). \tag{5.89}$$

For the proper homogeneous Lorentz transformations, we obtain from (5.88)

$$\begin{aligned}
\zeta_{\alpha(\lambda\mu)} &= -x_\gamma \frac{\partial}{\partial t^\alpha} (g^{\mu\mu} \delta_{\gamma\lambda} t^\mu - g^{\lambda\lambda} \delta_{\gamma\mu} t^\lambda) \\
&= -x_\gamma (g^{\mu\mu} \delta_{\gamma\lambda} \delta_{\mu\alpha} - g^{\lambda\lambda} \delta_{\gamma\mu} \delta_{\lambda\alpha}) \\
&= -g^{\mu\mu} \delta_{\mu\alpha} x_\lambda + g^{\lambda\lambda} \delta_{\lambda\alpha} x_\mu
\end{aligned} \tag{5.90}$$

for $(\lambda, \mu) \in S$. It should be clear that no summation over $\lambda$ and $\mu$ is intended above.

We are now in position to write down the conservation laws corresponding to a physical field which is described by a covariant vector field.

**5.3   Theorem**   A necessary condition for the Lagrangian density $L(x(t), \partial x(t))$ of a covariant vector field to be absolutely invariant under the ten-parameter Poincaré group (5.39) and (5.40) is that the following ten conservation laws hold true:

$$\frac{\partial}{\partial t^\alpha} H_\beta^{\ \alpha} = 0 \qquad (\beta = 0, \ldots, 3) \tag{5.91}$$

$$\frac{\partial}{\partial t^\alpha} \{-H_\lambda^{\ \alpha} g^{\mu\mu} t^\mu + H_\mu^{\ \alpha} g^{\lambda\lambda} t^\lambda - p^{\mu\alpha} g^{\mu\mu} x_\lambda + p^{\lambda\alpha} g^{\lambda\lambda} x_\mu\} = 0 \qquad (\lambda, \mu) \in S,$$

$$\tag{5.92}$$

where

$$H_\beta{}^\alpha \equiv -L\delta_\beta{}^\alpha + \frac{\partial L}{\partial(\partial x_\lambda/\partial t^\alpha)}\frac{\partial x_\lambda}{\partial t^\beta} \quad \text{and} \quad p^{\beta\alpha} \equiv \frac{\partial L}{\partial(\partial x_\beta/\partial t^\alpha)}. \tag{5.93}$$

*Proof*  (5.91) is a special case of Example 4.2. To prove (5.92), we observe from (4.36) that invariance under the proper homogeneous Lorentz group yields

$$\frac{\partial}{\partial t^\alpha}(-H_\beta{}^\alpha \tau^\beta_{(\lambda\mu)} + p^{\beta\alpha}\xi_{\beta(\lambda\mu)}) = 0 \qquad (\lambda,\mu) \in S, \tag{5.94}$$

where $\tau^\beta_{(\lambda\mu)}$ and $\xi_{\beta(\lambda\mu)}$ are defined by (5.43) and (5.90), respectively. Upon substitution of these quantities into (5.94), we immediately obtain (5.92).  □

In the present context the Hamiltonian $H_\beta{}^\alpha$ is sometimes known as the *energy–momentum tensor*. From (5.91) we infer that

$$\Pi_\beta \equiv \int H_\beta{}^0\, dt^1\, dt^2\, dt^3 \qquad (\beta = 0, \ldots, 3) \tag{5.95}$$

represent conserved quantities for the field. As in the previous section, we shall refer to $H_0{}^0$ as the energy density of the field, and $H_1{}^0$, $H_2{}^0$, and $H_3{}^0$ as the momentum densities. We leave the calculation of these quantities to the next few paragraphs where we discuss a particular field, namely, the electromagnetic field.

We shall now specialize the above results to the electromagnetic field in free space. The Hamiltonian for the field is, according to (5.93),

$$H_\beta{}^\alpha = -L\delta_\beta{}^\alpha + \frac{\partial A_\gamma}{\partial t^\beta}\frac{\partial L}{\partial(\partial A_\gamma/\partial t^\alpha)}.$$

A straightforward calculation shows that

$$\frac{\partial L}{\partial(\partial A_\gamma/\partial t^\alpha)} = F^{\gamma\alpha}, \tag{5.96}$$

if $L$ is given by (5.84). Hence the Hamiltonian is given by

$$H_\beta{}^\alpha = -L\delta_\beta{}^\alpha + F^{\gamma\alpha}\frac{\partial A_\gamma}{\partial t^\beta}. \tag{5.97}$$

The conservation laws can now be written down directly using (5.91) and (5.92). However, since $H_\beta{}^\alpha$ is not symmetric, it is common practice in electromagnetic theory to define another quantity $E_\beta{}^\alpha$ which is symmetric and whose divergence vanishes if the divergence of $H_\beta{}^\alpha$ vanishes. By way of motivation, it follows directly from (5.97) and the definition of $F_{\alpha\beta}$ that

$$H_\beta{}^\alpha = -L\delta_\beta{}^\alpha + F_{\beta\gamma} F^{\gamma\alpha} + \frac{\partial A_\beta}{\partial t^\gamma} F^{\gamma\alpha}. \tag{5.98}$$

We define the *energy–momentum tensor* $E_\beta{}^\alpha$ of the electromagnetic field by

$$E_\beta{}^\alpha = -L\delta_\beta{}^\alpha + F_{\beta\gamma} F^{\gamma\alpha}, \tag{5.99}$$

so that from (5.97) the relationship between this energy–momentum tensor and the Hamiltonian is

$$E_\beta{}^\alpha = H_\beta{}^\alpha - \frac{\partial A_\beta}{\partial t^\gamma} F^{\gamma\alpha}. \tag{5.100}$$

Now, it is clear that $E_\beta{}^\alpha = E_\alpha{}^\beta$, so that it is symmetric; moreover, to prove that it is divergence free, i.e.,

$$\frac{\partial E_\beta{}^\alpha}{\partial t^\alpha} = 0, \tag{5.101}$$

whenever $H_\beta{}^\alpha$ is divergence free, we differentiate (5.101) with respect to $t^\alpha$ to obtain

$$\frac{\partial E_\beta{}^\alpha}{\partial t^\alpha} = \frac{\partial H_\beta{}^\alpha}{\partial t^\alpha} - \frac{\partial A_\beta}{\partial t^\gamma}\frac{\partial F^{\gamma\alpha}}{\partial t^\alpha} - \frac{\partial^2 A_\beta}{\partial t^\alpha \, \partial t^\gamma} F^{\gamma\alpha}.$$

The first term on the right is zero by hypothesis, and the second term vanishes since $(\partial/\partial t^\alpha)F^{\gamma\alpha} = 0$ coincides with the Euler–Lagrange equations for the field. The third term vanishes since $F^{\gamma\alpha}$ is antisymmetric, i.e., $F^{\gamma\alpha} = -F^{\alpha\gamma}$. Hence, our assertion (5.101) is valid.

Rewriting (5.99) using the definition of the Lagrangian $L$, we get

$$E_\beta{}^\alpha = \tfrac{1}{4} F_{\gamma\delta} F^{\gamma\delta}\delta_\beta{}^\alpha + F_{\beta\gamma} F^{\gamma\alpha}. \tag{5.102}$$

Since $E_\beta{}^\alpha$ is divergence free by (5.101), it follows from the remarks in Section 4.4 that the time-invariants are the quantities

$$\Pi_\beta = \int E_\beta{}^0 \, dt^1 \, dt^2 \, dt^3.$$

In terms of the observable quantities $\mathbf{E}$ and $\mathbf{B}$, the energy–momentum tensor $E_\beta{}^\alpha$ is

$$E_\beta{}^\alpha = -\tfrac{1}{2}(\mathbf{E}^2 - \mathbf{B}^2)\delta_\beta{}^\alpha + F_{\beta\gamma} F^{\gamma\alpha}.$$

Therefore

$$E_0{}^0 = -\tfrac{1}{2}(\mathbf{E}^2 - \mathbf{B}^2) + F_{0\gamma} F^{\gamma 0}$$
$$= \tfrac{1}{2}(\mathbf{E}^2 + \mathbf{B}^2),$$

which is the energy density of the field. Consequently,

$$\Pi_0 = \int (\tfrac{1}{2}\mathbf{E}^2 + \tfrac{1}{2}\mathbf{B}^2)\, dt^1\, dt^2\, dt^3$$

is independent of time, and so we obtain conservation of energy. Computing the other conserved quantities, we obtain

$$E_1{}^0 = F_{1\gamma}F^{\gamma 0} = -B_3 E_2 + B_2 E_3,$$
$$E_2{}^0 = F_{2\gamma}F^{\gamma 0} = B_3 E_1 - B_1 E_3,$$
$$E_3{}^0 = F_{3\gamma}F^{\gamma 0} = B_1 E_2 - B_2 E_1.$$

We therefore observe that $-E_1{}^0$, $-E_2{}^0$, and $-E_3{}^0$ are components of the Poynting vector† $\mathbf{S} = \mathbf{E} \times \mathbf{B}$ which was defined in Section 5.7. The remaining components $E_\beta{}^\alpha(\alpha, \beta = 1, 2, 3)$ are the components of the well-known Maxwell stress tensor.

We leave as an exercise the calculation of the conservation laws (5.92) which arise from the invariance of the action integral under the homogeneous Lorentz transformations (see Rohrlich [1]).

## EXERCISES

**5-1**  Prove that the set of all general Lorentz transformations form a group under composition.

**5-2**  A vector $t = (t^0, t^1, t^2, t^3)^{\mathrm{T}}$ is called *positive timelike* if $t^0 > 0$ and $(-t^0)^2 + (t^1)^2 + (t^2)^2 + (t^3)^2 < 0$. Prove that a general Lorentz transformation satisfies the condition $a_0{}^0 \geq 1$ if, and only if, it maps positive timelike vectors to positive timelike vectors.

[*Hint*: Use the Cauchy–Schwarz inequality.]

**5-3**  Prove that the set of all proper Lorentz transformations form a group.

**5-4**  Let $T_\lambda^{\alpha\beta}$ be a tensor of type $\binom{2}{1}$. Show that "contraction on two upper indices" does not result in a tensor. In other words, show that the components $C_\lambda \equiv T_\lambda^{\alpha\alpha}$ do not form the components of a tensor. However, prove that $A^\beta \equiv T_\alpha^{\alpha\beta}$ is a tensor.

**5-5**  Let $\phi = \phi(t^0, \ldots, t^3)$ be a scalar function. Show that the partial derivatives $\partial\phi/\partial t^\alpha$ form the components of a covariant tensor.

**5-6**  Let $A_\alpha$ and $B^\alpha$ be convariant and contravariant tensors, respectively. Show that $A_\alpha B^\alpha$ is a scalar.

† Here, S has units of energy per unit volume.

**5-7**   The trace of a type $\binom{1}{1}$ tensor $T_j{}^i$ is defined to be $T_i{}^i$. Show that the trace is a scalar.

**5-8**   In $R^3$ the components $A_j{}^i$ of a tensor $A$ of type $\binom{1}{1}$ are given by

$$A_j{}^i = \begin{bmatrix} 2 & 0 & 1 \\ 0 & 3 & -1 \\ 1 & -1 & 0 \end{bmatrix}$$

in a basis $e_1, e_2, e_3$ and dual basis $\varepsilon^1, \varepsilon^2, \varepsilon^3$. Let $\bar{e}_1 = e_1 + e_2, \bar{e}_2 = 2e_2$, and $\bar{e}_3 = -e_2 + e_3$ be a new basis and let $\bar{\varepsilon}^k, k = 1, 2, 3$ be the new dual basis.
   (a)   Find the components $\bar{A}_j{}^i$ in the new basis.
   (b)   Find an expression for the $\bar{\varepsilon}^k$ in terms of the original dual basis $\varepsilon^k, k = 1, 2, 3$.
   (c)   Let $v = -e_1 + 2e_2$ and $\tau = 5\varepsilon^1 - 2\varepsilon^2 + \varepsilon^3$. Evaluate $A(\tau, v)$.
   (d)   Redo part (c) in the new basis and show that the result is the same.

**5-9**   Show that the Lagrangian defined by (5.53) satisfies the condition (5.59).

**5-10**   Verify that if the Lagrangian $L$ is given by (5.28), then

$$\frac{\partial L}{\partial(\partial A_\alpha/\partial t^\beta)} = F^{\alpha\beta}.$$

**5-11**   (a)   Derive the Euler–Lagrange equations corresponding to the Lagrangian density

$$L = \sum_{\alpha=0}^{3} \varepsilon_\alpha \left(\frac{\partial \phi}{\partial t^\alpha} - eA_\alpha\right)^2 + m^2\phi^2 + \sum_{\alpha,\beta=0}^{3} \varepsilon_\alpha \varepsilon_\beta \left(\frac{\partial A_\alpha}{\partial t^\beta}\right)^2 + m \sum_{\alpha=0}^{3} \varepsilon_\alpha A_\alpha{}^2,$$

where the field functions are $\phi, A_0, A_1, A_2$, and $A_3$, and

$$\varepsilon_\alpha = \begin{cases} 1 & \text{if } \alpha = 0, \\ -1 & \text{if } \alpha = 1, 2, 3. \end{cases}$$

$e$ and $m$ are constants.
   (b)   Assuming that $\phi$ is a scalar function and the $A_\alpha$ are components of a covariant vector, use the fact that $\int L\, dt^0 \cdots dt^3$ is Lorentz invariant to derive the conservation laws of the field defined by $L$.

**5-12**   Suppose the scalar field $\phi = \phi(\bar{t}, \bar{x}^1, \bar{x}^2, \bar{x}^3)$ satisfies the wave equation

$$\sum_{i=1}^{3} \frac{\partial^2 \phi}{\partial \bar{x}^{i\,2}} - \frac{1}{c^2}\frac{\partial^2 \phi}{\partial \bar{t}^2} = 0$$

in the $(\bar{t}, \bar{x}^1, \bar{x}^2, \bar{x}^3)$ coordinate system. If

$$x^k = \bar{x}^k - v^k\bar{t} \qquad t = \bar{t},$$

show that in the coordinate system $(t, x^1, x^2, x^3)$ the wave equation becomes

$$\nabla^2 \phi - \frac{1}{c^2} \frac{\partial^2 \phi}{\partial t^2} - \frac{2}{c^2} v^k \frac{\partial}{\partial x^k} \left( \frac{\partial \phi}{\partial t} \right) - \frac{1}{c^2} v^k \frac{\partial v^l}{\partial x^k} \frac{\partial \phi}{\partial x^l} = 0.$$

Hence the wave equation is not form invariant under Galilean transformations.

**5-13**   If $A_j$ is a tensor of type $\binom{0}{1}$, show that

$$F_{kj} \equiv \frac{\partial A_j}{\partial t^k} - \frac{\partial A_k}{\partial t^j}$$

is a tensor of type $\binom{0}{2}$.

# CHAPTER 6

# Second-Order Variation Problems

## 6.1 THE EULER–LAGRANGE EQUATIONS

Problems in the calculus of variations whose Lagrange function involves higher-order derivatives have received considerable attention ever since the origins of the subject in the early eighteenth century. At various times it has been maintained that such problems do not merit a separate theoretical treatment since they may be easily reduced to a problem of Lagrange involving constraints (see McFarlan [1]); however, such an objection applies only in the case of nonparameter-invariant problems, i.e., variational problems which are not invariant under arbitrary transformations of the independent variables (Grässer [1]). The continued interest in second-order problems lies in the fact that higher-order problems can be applied, with varying degrees of success, to other branches of mathematics and to physics. From the point of view of physical applications, such problems enjoy considerable application in relativity and continuum mechanics. For example, in general relativity the Lagrange function which gives rise to the Einstein gravitational field equations is $L = R\sqrt{-g}$, where $R$ is the scalar curvature and $g$ is the determinant of the metric tensor. The scalar curvature $R$ inherently contains second derivatives of the components of the metric tensor.

Also, in continuum mechanics the governing differential equations are often of fourth order and consequently the associated Lagrangian is of second order. Efforts have also been made to establish a "generalized mechanics" and a "generalized electrodynamics" by including higher-order derivatives in the Lagrangian, but at the present time it would be premature to comment on the significance of these extensions of the classical theory.

In the single integral case, the variational integral for the second-order problem is

$$J = \int_a^b L(t, x(t), \dot{x}(t), \ddot{x}(t))\, dt, \tag{6.1}$$

where $x(t) = (x^1(t), \ldots, x^n(t)) \in C_n{}^4(a, b)$, $\dot{x}(t) = dx/dt$, and $\ddot{x}(t) = d^2x/dt^2$. The Lagrangian $L$ is assumed to be of class $C^4$ in each of its $3n + 1$ variables. In the multiple integral case, the second-order problem is defined by the integral

$$J = \int_D L(t, x(t), \partial x(t), \partial^2 x(t))\, dt^1 \cdots dt^m, \tag{6.2}$$

where $t = (t^1, \ldots, t^m)$, $x(t) = (x^1(t), \ldots, x^n(t)) \in C_n{}^4(D)$, and $\partial x(t)$ and $\partial^2 x(t)$ denote the collection of first and second partial derivatives

$$\dot{x}_\alpha{}^k = \frac{\partial x^k}{\partial t^\alpha}, \qquad \ddot{x}_{\alpha\beta}^k = \frac{\partial^2 x}{\partial t^\beta\, \partial t^\alpha},$$

respectively. In this case, the Lagrangian depends upon $m + n + mn + m^2n$ variables, so that we assume that $\ddot{x}_{\alpha\beta}^k$ occurs as well as $\ddot{x}_{\beta\alpha}^k$. This is a notational convenience; in practice, only one of these second derivatives will occur since mixed partials are equal (because of our assumption that $x(t) \in C^4(D)$).

The derivation of the Euler–Lagrange equations corresponding to (6.1) and (6.2) proceeds in manner similar to that in Chapter 1 for first-order problems. We shall briefly carry out this calculation for the more general multiple integral problem, leaving the single integral calculation as an exercise.

Therefore, let us consider the variational problem

$$J(x) = \int_D L(t, x(t), \partial x(t), \partial^2 x(t))\, dt^1 \cdots dt^m \to \min, \tag{6.3}$$

where $x(t) \in C_n{}^4(D)$ satisfies the additional boundary conditions

$$x^k(t) = \Phi^k(t), \qquad \dot{x}_\alpha{}^k(t) = \Psi_\alpha{}^k(t), \qquad t \in \mathrm{Bd}\, D, \tag{6.4}$$

where $\Phi^k$ and $\Psi_\alpha{}^k$ are given, fixed functions on the boundary of $D$. To compute the first variation of $J$ we embed $x(t)$ in a one-parameter family of surfaces

$$\bar{x}(t) = x(t) + \varepsilon\eta(t),$$

where $\eta(t) \in C_n{}^4(D)$ and satisfies the boundary conditions

$$\eta^k(t) = 0, \qquad \dot{\eta}_\alpha{}^k(t) = 0, \qquad t \in \text{Bd } D. \tag{6.5}$$

Then $\bar{x}(t)$ is an admissible surface and

$$\delta J(x, \eta) = \frac{\partial}{\partial\varepsilon} \int_D L(t, x(t) + \varepsilon\eta(t), \partial(x(t) + \varepsilon\eta(t)),$$

$$\partial^2(x(t) + \varepsilon\eta(t))) \, dt^1 \cdots dt^m \Big|_{\varepsilon=0}$$

$$= \int_D \left( \frac{\partial L}{\partial x^k}\eta^k + \frac{\partial L}{\partial \dot{x}_\alpha{}^k}\dot{\eta}_\alpha{}^k + \frac{\partial L}{\partial \ddot{x}_{\alpha\beta}^k}\ddot{\eta}_{\alpha\beta}^k \right) dt^1 \cdots dt^m.$$

When the second two terms in the integrand are integrated by parts we get

$$\delta J(x, \eta) = \int_D \left( \frac{\partial L}{\partial x^k} - \frac{\partial}{\partial t^\alpha}\frac{\partial L}{\partial \dot{x}_\alpha{}^k} + \frac{\partial^2}{\partial t^\alpha \partial t^\beta}\frac{\partial L}{\partial \ddot{x}_{\alpha\beta}^k} \right)\eta^k \, dt^1 \cdots dt^m$$

$$+ \int_D \left[ \frac{\partial}{\partial t^\alpha}\left( \frac{\partial L}{\partial \dot{x}_\alpha{}^k}\dot{\eta}^k \right) - \frac{\partial}{\partial t^\beta}\frac{\partial L}{\partial \ddot{x}_{\alpha\beta}^k}\eta^k \right) + \frac{\partial}{\partial t^\beta}\left( \frac{\partial L}{\partial \ddot{x}_{\alpha\beta}^k}\dot{\eta}_\alpha{}^k \right) \right] dt^1 \cdots dt^m.$$

The second integral, being the integral of a divergence, can be transformed via the Divergence Theorem to

$$\int_{\text{Bd } D} \left[ \left( \frac{\partial L}{\partial \dot{x}_\alpha{}^k} - \frac{\partial}{\partial t^\beta}\frac{\partial L}{\partial \ddot{x}_{\alpha\beta}^k} \right)\eta^k + \frac{\partial L}{\partial \ddot{x}_{\beta\alpha}^k}\dot{\eta}_\beta{}^k \right] \cos(n, t^\alpha) \, d\sigma.$$

From the assumed boundary conditions (6.5) on $\eta(t)$, we see that this surface integral vanishes and thus we have:

**6.1   Lemma**   A necessary condition for $x = x(t)$ to minimize the variational problem defined by (6.3) and (6.4) is that

$$\int_D \left( \frac{\partial L}{\partial x^k} - \frac{\partial}{\partial t^\alpha}\frac{\partial L}{\partial \dot{x}_\alpha{}^k} + \frac{\partial^2}{\partial t^\alpha \partial t^\beta}\frac{\partial L}{\partial \ddot{x}_{\alpha\beta}^k} \right)\eta^k \, dt^1 \cdots dt^m = 0$$

for all $\eta \in C_n{}^4(D)$ satisfying the conditions (6.5).   □

An application of the Fundamental Lemma of the calculus of variations will give the desired result, namely:

**6.1   Theorem**   A necessary condition for $x = x(t)$ to minimize the variational problem defined by (6.3) and (6.4) is that its components satisfy the $n$ partial differential equations

$$E_k^{(2)} \equiv \frac{\partial L}{\partial x^k} - \frac{\partial}{\partial t^\alpha} \frac{\partial L}{\partial \dot{x}_\alpha^k} + \frac{\partial^2}{\partial t^\alpha \, \partial t^\beta} \frac{\partial L}{\partial \ddot{x}_{\alpha\beta}^k}$$

$$= 0 \qquad (k = 1, \ldots, n). \qquad \square \qquad\qquad (6.6)$$

Equations (6.6) are the *Euler–Lagrange equations* for the second-order multiple integral problem. They are a system of fourth-order nonlinear partial differential equations; as before, any solution of (6.6) will be called an *extremal*. $E_k^{(2)}$ is our notation for the left-hand side of (6.6), the Euler expressions.

Using these results for the multiple integral problem, it is easy to derive the corresponding results for the single integral case by taking $m = 1$. To this end, let us consider the problem

$$J(x) = \int_a^b L(t, x, \dot{x}, \ddot{x}) \, dt \to \min, \qquad\qquad (6.7)$$

where $x \in C_n{}^4(a, b)$ and satisfies the boundary conditions

$$x(a) = A_1, \qquad x(b) = B_1, \qquad \dot{x}(a) = A_2, \qquad \dot{x}(b) = B_2. \qquad (6.8)$$

Taking $m = 1$ in Theorem 6.1 we obtain the following result for single integrals.

**6.2   Theorem**   A necessary condition for $x = x(t)$ to minimize the variation problem defined by (6.7) and (6.8) is that its components satisfy the $n$ fourth-order ordinary differential equations

$$\frac{\partial L}{\partial x^k} - \frac{d}{dt} \frac{\partial L}{\partial \dot{x}^k} + \frac{d^2}{dt^2} \frac{\partial L}{\partial \ddot{x}^k} = 0 \qquad (k = 1, \ldots, n). \qquad \square \qquad (6.9)$$

Therefore, Eqs. (6.9) are the *Euler–Lagrange equations* for the second-order single integral problem.

**6.1   Example**   In investigating the vibrations of a fixed wedge of constant thickness (Fig. 7), we must examine (see Elsgolc [1]) extremals of the functional

$$J(x) = \int_0^1 (at^3\ddot{x}^2 - btx^2) \, dt,$$

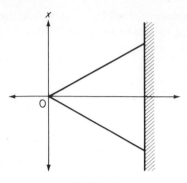

**FIGURE 7**

where $a, b > 0$. In this case the extremals are, according to (6.9), solutions of the linear fourth-order differential equation

$$at^3 \frac{d^4x}{dt^4} + 6at^2 \frac{d^3x}{dt^3} + 6at \frac{d^2x}{dt^2} - btx = 0,$$

which is the Euler–Lagrange equation for this problem.

**6.2  Example**  (*Vibrating Rod*)  We consider an elastic rod of length $l$ whose position of equilibrium lies along the $x$ axis. The deflection from equilibrium is $u(t, x)$, where $t$ is time and $x$ is position. The kinetic energy of the rod is

$$T = \int_0^l \frac{1}{2} \rho u_t^2 \, dx,$$

where $\rho$ is its density. Assuming the rod will not stretch, the potential energy is then proportional to the square of its curvature, and so

$$U = \int_0^l \frac{1}{2} k \left[ \frac{u_{xx}}{(1 + u_x^2)^{3/2}} \right]^2 dx$$

$$\cong \frac{1}{2} \int_0^l k u_{xx}^2 \, dx,$$

the latter approximation being due to an assumption of small deflections from equilibrium. The variation integral is then, according to Hamilton's principle for systems of material points,

$$J = \int_{t_0}^{t_1} (T - U) \, dt$$

$$= \int_{t_0}^{t_1} \int_0^l (\tfrac{1}{2}\rho u_t^2 - \tfrac{1}{2}k u_{xx}^2) \, dx \, dt.$$

The Lagrangian here involves second derivatives, and making the asso-
ciations $t \leftrightarrow t^1$, $x \leftrightarrow t^2$, $u(t, x) \leftrightarrow x^1(t^1, t^2)$, to conform with our earlier
notation, we note that the Euler–Lagrange equation becomes

$$\rho u_{tt} + k u_{xxxx} = 0,$$

which is the governing equation for the vibration of the rod.


## 6.2   INVARIANCE CRITERIA FOR SINGLE INTEGRALS

In this section we derive the fundamental invariance identities for second-
order problems in the single integral case. As in Chapter 2 we consider an
$r$-parameter family of transformations

$$\bar{t} = \phi(t, x, \varepsilon), \qquad \bar{x}^k = \psi^k(t, x, \varepsilon) \qquad (k = 1, \ldots, n) \qquad (6.10)$$

of $(t, x)$-space such that the mappings $\phi$ and $\psi^k$ are of class $C^3$ in each of
their arguments, and such that the family contains identity mapping:

$$\bar{t} = \phi(t, x, 0) = t, \qquad \bar{x}^k = \psi^k(t, x, 0) = x^k.$$

The coefficients of the linear terms in the $\varepsilon^s$ in the Taylor series expansion of
(6.10) about $\varepsilon = 0$ are the *infinitesimal generators*

$$\tau_s(t, x) \equiv \left.\frac{\partial \phi}{\partial \varepsilon^s}\right|_{\varepsilon = 0} \qquad \text{and} \qquad \xi_s^{\,k}(t, x) \equiv \left.\frac{\partial \psi^k}{\partial \varepsilon^s}\right|_{\varepsilon = 0}. \qquad (6.11)$$

As in Chapter 2, each function $x = x(t) \in C_n^{\,4}(t_0, t_1)$ gets mapped via (6.10)
into an $r$-parameter family of functions $\bar{x} = \bar{x}(\bar{t})$ in $(\bar{t}, \bar{x})$-space, provided $\varepsilon$
lies in a sufficiently small rectangle about the origin in $R^r$. Therefore, we may
state the following definition:

**6.1   Definition**   The fundamental integral (6.1) is invariant under the
$r$-parameter family of transformations (6.10) if and only if given any curve
$x: [a, b] \to R^n$ of class $C^4$ and $a \le t_1 < t_2 \le b$, we have

$$\int_{\bar{t}_1}^{\bar{t}_2} L\left(\bar{t}, \bar{x}(\bar{t}), \frac{d\bar{x}}{d\bar{t}}, \frac{d^2\bar{x}}{d\bar{t}^2}\right) d\bar{t} - \int_{t_1}^{t_2} L(t, x(t), \dot{x}(t), \ddot{x}(t))\, dt = o(\varepsilon) \qquad (6.12)$$

for all sufficiently small $\varepsilon$, where

$$\bar{t}_i = \phi(t_i, x(t_i), \varepsilon) \qquad (i = 1, 2).$$

The analog of the fundamental theorem, Theorem 2.1, for second-order
problems is the following result:

**6.3   Theorem**   If the fundamental integral (6.1) is invariant under the *r*-parameter family of transformations (6.10), then the Lagrangian *L* and its derivatives satisfy the *r* identity relations

$$\frac{\partial L}{\partial t}\tau_s + \frac{\partial L}{\partial x^k}\xi_s^{\ k} + \frac{\partial L}{\partial \dot{x}^k}\left(\frac{d\xi_s^{\ k}}{dt} - \dot{x}^k\frac{d\tau_s}{dt}\right),$$

$$+ \frac{\partial L}{\partial \ddot{x}^k}\left(\frac{d^2\xi_s^{\ k}}{dt^2} - 2\ddot{x}^k\frac{d\tau_s}{dt} - \dot{x}^k\frac{d^2\tau_s}{dt^2}\right) + L\frac{d\tau_s}{dt} = 0, \qquad (6.13)$$

where $s = 1, \ldots, r$ and $\tau_s$ and $\xi_s^{\ k}$ are defined by (6.11).

Comparing (6.13) with (2.15), we observe that the only difference is the presence of one additional term in the second-order case. Also, in the second-order case it is possible to generalize the transformations (6.10) by allowing them to depend upon the derivatives $\dot{x}$; such an analysis shows, although we choose not to present the argument here, that there is no change in the Eqs. (6.13). However, when the total derivatives of $\tau_s$ and $\xi_s^{\ k}$ are expanded in (6.13), extra terms will occur due to the presence of an additional variable. Finally, in Definition 6.1 we have defined invariance, or absolute invariance, instead of the more general divergence-invariance. The latter can be defined by including on the right-hand side of (6.12) the integral of exact derivative which is linear in the $\varepsilon^s$, i.e.,

$$\int_{t_1}^{t_2} \varepsilon^s \frac{dF_s}{dt}\,dt.$$

The resulting invariance identities (6.13) will then have a $dF_s/dt$ term on the right-hand side.

The proof of Theorem 6.3 follows along the same lines as the proof of Theorem 2.1—we differentiate (6.12) with respect to $\varepsilon^s$ and afterwards set $\varepsilon = 0$. Changing variables in the first integral in (6.12) and performing the differentiation, we obtain, after some simplification,

$$L(t, x, \dot{x}, \ddot{x})\left(\frac{\partial}{\partial\varepsilon^s}\frac{d\bar{t}}{dt}\right)_0 + \frac{\partial}{\partial\varepsilon^s}L\left(\bar{t}, \bar{x}(\bar{t}), \frac{d\bar{x}}{d\bar{t}}, \frac{d^2\bar{x}}{d\bar{t}^2}\right)_0 = 0,$$

which becomes

$$L\frac{d\tau_s}{dt} + \frac{\partial L}{\partial t}\tau_s + \frac{\partial L}{\partial x^k}\xi_s^{\ k} + \frac{\partial L}{\partial \dot{x}^k}\left(\frac{\partial}{\partial\varepsilon^s}\frac{d\bar{x}^k}{d\bar{t}}\right)_0 + \frac{\partial L}{\partial \ddot{x}^k}\left(\frac{\partial}{\partial\varepsilon^s}\frac{d^2\bar{x}^k}{d\bar{t}^2}\right)_0 = 0, \quad (6.14)$$

where we have used (2.18) and (2.16). From (2.20) we can evaluate the fourth term, and the last term can be computed as follows. Differentiating both sides of (2.8) with respect to *t* we obtain

$$\frac{d\bar{x}^k}{d\bar{t}}\frac{d\bar{t}}{dt} = \frac{d\psi^k}{dt}.$$

Differentiating again, we get

$$\frac{d\bar{x}^k}{d\bar{t}}\frac{d^2\bar{t}}{dt^2} + \frac{d^2\bar{x}^k}{d\bar{t}^2}\left(\frac{d\bar{t}}{dt}\right)^2 = \frac{d^2\psi^k}{dt^2}$$

Now, differentiating with respect to $\varepsilon^s$ and subsequently setting $\varepsilon = 0$ gives

$$\frac{d^2\tau_s}{dt^2}\ddot{x}^k + \left(\frac{\partial}{\partial\varepsilon^s}\frac{d^2\bar{x}^k}{d\bar{t}^2}\right)_0 + 2\dot{x}^k\frac{d\tau_s}{dt} = \frac{d^2\zeta_s{}^k}{dt^2}, \tag{6.15}$$

where we have used the facts that

$$\left(\frac{d^2\bar{t}}{dt^2}\right)_0 = 0, \qquad \left(\frac{\partial}{\partial\varepsilon^s}\frac{d^2\bar{t}}{dt^2}\right)_0 = \frac{d^2\tau_s}{dt^2}, \qquad \left(\frac{d^2\bar{x}^k}{d\bar{t}^2}\right)_0 = \ddot{x}^k,$$

$$\left(\frac{\partial}{\partial\varepsilon^s}\frac{d^2\psi^k}{dt^2}\right)_0 = \frac{d^2\zeta_s{}^k}{dt^2},$$

which follow from (6.10), along with (2.16). Solving for the $((\partial/\partial\varepsilon^s)(d^2\bar{x}^k/d\bar{t}^2))_0$ term in (6.15) and substituting the result along with (2.20) into (6.14), we obtain the identities (6.13) and thus complete the proof. □

Immediately, we can prove Noether's theorem for second-order problems.

**6.4  Theorem**  (*Noether*)  If the fundamental integral (6.1) is invariant under the $r$-parameter family of transformations (6.10), then the following $r$ identities are valid:

$$-E_k^{(2)}(\zeta_s{}^k - \dot{x}^k\tau_s) = \frac{d}{dt}\left[L\tau_s + \left(\frac{\partial L}{\partial\dot{x}^k} - \frac{d}{dt}\frac{\partial L}{\partial\ddot{x}^k}\right)(\zeta_s{}^k - \dot{x}^k\tau_s)\right.$$

$$\left. + \frac{\partial L}{\partial\ddot{x}^k}\frac{d}{dt}(\zeta_s{}^k - \dot{x}^k\tau_s)\right], \tag{6.16}$$

where

$$E_k^{(2)} \equiv \frac{\partial L}{\partial x^k} - \frac{d}{dt}\frac{\partial L}{\partial\dot{x}^k} + \frac{d^2}{dt^2}\frac{\partial L}{\partial\ddot{x}^k} \tag{6.17}$$

are the Euler–Lagrange expressions for the second-order problem.

*Proof*  The essential idea is to notice that the *Noether identities* (6.16) involve at least fourth-order derivatives of the functions $x^k(t)$ through the presence of the expression $d^2/dt^2\,(\partial L/\partial\ddot{x}^k)$, whereas the invariance identities (6.13) involve at most second derivatives. The introduction of the higher derivatives in (6.13) can be accomplished by sustituting the following expressions into (6.13):

$$\frac{\partial L}{\partial t}\tau_s = \frac{dL}{dt}\tau_s - \dot{x}^k\frac{\partial L}{\partial x^k}\tau_s - \ddot{x}^k\frac{\partial L}{\partial \dot{x}^k}\tau_s - \dddot{x}^k\frac{\partial L}{\partial \ddot{x}^k}\tau_s,$$

$$-\dot{x}^k\frac{\partial L}{\partial \ddot{x}^k}\frac{d\tau_s}{dt} = -\frac{d}{dt}\left(\dot{x}^k\frac{\partial L}{\partial \ddot{x}^k}\tau_s\right) + \ddot{x}^k\frac{\partial L}{\partial \ddot{x}^k}\tau_s + \dot{x}^k\frac{d}{dt}\frac{\partial L}{\partial \ddot{x}^k}\tau_s,$$

$$-\dot{x}^k\frac{\partial L}{\partial \ddot{x}^k}\frac{d^2\tau_s}{dt^2} = -\frac{d}{dt}\left(\dot{x}^k\frac{\partial L}{\partial \ddot{x}^k}\frac{d\tau_s}{dt}\right) + \ddot{x}^k\frac{\partial L}{\partial \ddot{x}^k}\frac{d\tau_s}{dt} + \frac{d}{dt}\left(\dot{x}^k\frac{d}{dt}\frac{\partial L}{\partial \ddot{x}^k}\tau_s\right)$$

$$-\ddot{x}^k\frac{d}{dt}\frac{\partial L}{\partial \ddot{x}^k}\tau_s - \dot{x}^k\frac{d^2}{dt^2}\frac{\partial L}{\partial \ddot{x}^k}\tau_s,$$

$$\frac{\partial L}{\partial \dot{x}^k}\frac{d\xi_s^{\,k}}{dt} = \frac{d}{dt}\left(\frac{\partial L}{\partial \dot{x}^k}\xi_s^{\,k}\right) - \frac{d}{dt}\frac{\partial L}{\partial \dot{x}^k}\xi_s^{\,k},$$

$$\frac{\partial L}{\partial \ddot{x}^k}\frac{d^2\xi_s^{\,k}}{dt^2} = \frac{d}{dt}\left(\frac{\partial L}{\partial \ddot{x}^k}\frac{d\xi_s^{\,k}}{dt}\right) + \frac{d^2}{dt^2}\frac{\partial L}{\partial \ddot{x}^k}\xi_s^{\,k} - \frac{d}{dt}\left(\frac{d}{dt}\frac{\partial L}{\partial \ddot{x}^k}\xi_s^{\,k}\right).$$

When simplifications are made, the Noether identities follow, and the proof is complete.   □

A simple corollary on conservation laws is provided by the following:

**6.1   Corollary**   If the fundamental integral (6.1) is invariant under the $r$-parameter family of transformations (6.10), and if $x(t)$ is an extremal, then

$$L\tau_s + \left(\frac{\partial L}{\partial \dot{x}^k} - \frac{d}{dt}\frac{\partial L}{\partial \ddot{x}^k}\right)(\xi_s^{\,k} - \dot{x}^k\tau_s) + \frac{\partial L}{\partial \ddot{x}^k}\frac{d}{dt}(\xi_s^{\,k} - \dot{x}^k\tau_s) = \text{constant}, \quad (6.18)$$

where $s = 1, \ldots, r$.   □

Equations (6.18) represent conservation laws or first integrals of the equations of motion (6.9). For example, if (6.1) is invariant under a time translation

$$\bar{t} = t + \varepsilon, \qquad \bar{x}^k = x^k,$$

then $\tau_s = 1$ and $\xi_s^{\,k} = 0$ (there is only one parameter), and Eq. (6.18) becomes

$$L + \dot{x}^k\frac{d}{dt}\frac{\partial L}{\partial \ddot{x}^k} - \dot{x}^k\frac{\partial L}{\partial \dot{x}^k} - \ddot{x}^k\frac{\partial L}{\partial \ddot{x}^k} = \text{constant}. \qquad (6.19)$$

Following the first-order analogy, we might consider the left-hand side of (6.19) to be a Hamiltonian for the second-order problem since it is the Hamiltonian that is conserved in the first-order case under time translations. Equation (6.19), then, represents a generalized statement of conservation of energy. If the fundamental integral is invariant under the spatial translations ($n$ parameters)

$$\bar{t} = t, \qquad \bar{x}^k = x^k + \varepsilon^k,$$

then $\tau_s = 0$ and $\xi_s{}^k = \delta_s{}^k$, the Kronecker delta. Equations (6.18) in this case become

$$\frac{\partial L}{\partial \dot{x}^k} - \frac{d}{dt}\frac{\partial L}{\partial \ddot{x}^k} = \text{constant} \qquad (k = 1, \ldots, n).$$

Again considering the analogy with first-order problems, there is ample reason to define the *canonical momenta* of a second-order problem to be

$$P_k = \frac{\partial L}{\partial \dot{x}^k} - \frac{d}{dt} R_k,$$

where

$$R_k = \frac{\partial L}{\partial \ddot{x}^k}.$$

Indeed, these are the relevant quantities for a second-order canonical formalism (see for example, Grässer [1], for a comprehensive discussion of this point). We summarize the preceding results in the following:

**6.2  Corollary**  The Euler–Lagrange equations (6.9) of the functional

$$J = \int_a^b L(t, x, \dot{x}, \ddot{x}) \, dt$$

have first integrals

$$\frac{\partial L}{\partial \dot{x}^k} - \frac{d}{dt}\frac{\partial L}{\partial \ddot{x}^k} = \text{constant} \qquad (k = 1, \ldots, n)$$

if the Lagrangian $L$ does not depend explicitly upon the $x^k$, and first integral

$$L + \dot{x}^k \frac{d}{dt}\frac{\partial L}{\partial \ddot{x}^k} - \dot{x}^k \frac{\partial L}{\partial \dot{x}^k} - \ddot{x}^k \frac{\partial L}{\partial \ddot{x}^k} = \text{constant}$$

if $L$ does not depend explicitly on $t$.   □


## 6.3  MULTIPLE INTEGRALS

We now carry out the program of the previous section for the multiple integral case, which seems to be the more important for applications. Therefore, we consider the fundamental integral (6.2) and an $r$-parameter family of transformations

$$\bar{t}^\alpha = \phi^\alpha(t, x, \varepsilon), \qquad \bar{x}^k = \psi^k(t, x, \varepsilon), \qquad (6.20)$$

where $\varepsilon = (\varepsilon^1, \ldots, \varepsilon^r)$ and $\alpha = 1, \ldots, m$; $k = 1, \ldots, n$, and which are of class $C^2$ in each of their $m + n + r$ arguments. Moreover, it is assumed that

$$\phi^\alpha(t, x, 0) = t^\alpha \qquad \text{and} \qquad \psi^k(t, x, 0) = x^k.$$

The infinitesimal generators of the transformations (6.20) are defined

$$\tau_s^\alpha(t, x) \equiv \left. \frac{\partial \phi^\alpha}{\partial \varepsilon^s} \right|_{\varepsilon = 0}, \qquad \xi_s^k(t, x) \equiv \left. \frac{\partial \psi^k}{\partial \varepsilon^s} \right|_{\varepsilon = 0}. \tag{6.21}$$

It can be shown that for $\varepsilon$ sufficiently small, the transformations (6.20) map a surface $x = x(t)$, $t \in D$, to another surface $\bar{x} = \bar{x}(\bar{t})$ in $(\bar{t}, \bar{x})$-space, and so we are permitted to make the following definition.

**6.2   Definition**   The fundamental integral (6.2) is (absolutely) invariant under the $r$-parameter family of transformations (6.20) if and only if for any $x \in C_n{}^4(D)$ and for any compact subset $H \subseteq D$, we have

$$\int_{\bar{H}} L(\bar{t}, \bar{x}(\bar{t}), \partial \bar{x}(\bar{t}), \partial^2 \bar{x}(\bar{t})) \, d\bar{t}^1 \cdots d\bar{t}^m$$

$$- \int_H L(t, x(t), \partial x(t), \partial^2 x(t)) \, dt^1 \cdots dt^m = o(\varepsilon) \tag{6.22}$$

for all $\varepsilon$ sufficiently small.

We are now able to state and prove the fundamental invariance theorem which is the multiple integral analog of Theorem 6.3.

**6.5   Theorem**   If the fundamental integral defined by (6.2) is invariant under the $r$-parameter family of transformations (6.20), then the Lagrangian $L$ and its derivatives satisfy the $r$ identities

$$\frac{\partial L}{\partial t^\alpha} \tau_s^\alpha + \frac{\partial L}{\partial x^k} \xi_s^k + \frac{\partial L}{\partial \dot{x}_\alpha^k} \left( \frac{d\xi_s^k}{dt^\alpha} - \dot{x}_\gamma^k \frac{d\tau_s^\gamma}{dt^\alpha} \right)$$

$$+ \frac{\partial^2 L}{\partial \ddot{x}_{\alpha\beta}^k} \left( \frac{d^2 \xi_s^k}{dt^\alpha \, dt^\beta} - \ddot{x}_{\beta\gamma}^k \frac{d\tau_s^\gamma}{dt^\alpha} - \ddot{x}_{\gamma\alpha}^k \frac{d\tau_s^\gamma}{dt^\beta} - \dot{x}_\gamma^k \frac{d^2 \tau_s^\gamma}{dt^\alpha \, dt^\beta} \right) + L \frac{d\tau_s^\alpha}{dt^\alpha} = 0 \tag{6.23}$$

for $s = 1, \ldots, r$, where $\tau_s^\alpha$ and $\xi_s^k$ are given by (6.21).

*Proof*   Differentiating (6.22) with respect to $\varepsilon^s$ and afterwards setting $\varepsilon = 0$ gives, as we have indicated in Chapter 4,

$$\frac{\partial L}{\partial t^\alpha} \tau_s^\alpha + \frac{\partial L}{\partial x^k} \xi_s^k + \frac{\partial L}{\partial \dot{x}_\alpha^k} \left( \frac{\partial}{\partial \varepsilon^s} \frac{\partial \bar{x}^k}{\partial \bar{t}^\alpha} \right)_0 + \frac{\partial^2 L}{\partial \ddot{x}_{\alpha\beta}^k} \left( \frac{\partial}{\partial \varepsilon^s} \frac{\partial^2 \bar{x}^k}{\partial \bar{t}^\alpha \, \partial \bar{t}^\beta} \right)_0 + L \frac{d\tau_s^\alpha}{dt^\alpha} = 0. \tag{6.24}$$

To obtain this, of course, we have changed variables in the transformed integral and eliminated the integrals by appealing to the arbitrary nature of the region $H$. It remains now to compute the two terms evaluated at zero in (6.24). This can be accomplished by differentiating (4.5) with respect to $t^\beta$ to obtain

$$\frac{\partial \bar{x}^k}{\partial \bar{t}^\alpha} \frac{\partial \bar{t}^\alpha}{\partial t^\beta} = \frac{\partial \bar{x}^k}{\partial t^\beta}. \tag{6.25}$$

Differentiation of (6.25) with respect to $\varepsilon^s$ then yields, after simplification and evaluation at $\varepsilon = 0$,

$$\left( \frac{\partial}{\partial \varepsilon^s} \frac{\partial \bar{x}^k}{\partial \bar{t}^\beta} \right)_0 = \frac{d\xi_s^k}{dt^\beta} - \dot{x}_\alpha^k \frac{d\tau_s^\alpha}{dt^\beta}. \tag{6.26}$$

Now we differentiate (6.25) with respect to $t^\gamma$ to obtain

$$\frac{\partial^2 \bar{x}^k}{\partial \bar{t}^\delta \partial \bar{t}^\alpha} \frac{\partial \bar{t}^\delta}{\partial t^\gamma} \frac{\partial \bar{t}^\alpha}{\partial t^\beta} + \frac{\partial \bar{x}^k}{\partial \bar{t}^\alpha} \frac{\partial^2 \bar{t}^\alpha}{\partial t^\gamma \partial t^\beta} = \frac{\partial^2 \bar{x}^k}{\partial t^\gamma \partial t^\beta}.$$

Then, differentiation of this expression with respect to $\varepsilon^s$ and evaluation at $\varepsilon = 0$ gives

$$\left( \frac{\partial}{\partial \varepsilon^s} \frac{\partial^2 \bar{x}^k}{\partial \bar{t}^\gamma \partial \bar{t}^\beta} \right)_0 = \frac{d^2 \xi_s^k}{dt^\gamma dt^\beta} - \ddot{x}_{\delta\beta}^k \frac{d\tau_s^\delta}{dt^\gamma} - \ddot{x}_{\gamma\alpha}^k \frac{d\tau_s^\alpha}{dt^\beta} - \dot{x}_\alpha^k \frac{d^2 \tau_s^\alpha}{dt^\gamma dt^\beta}, \tag{6.27}$$

where we have used the expressions

$$\left( \frac{\partial}{\partial \varepsilon^s} \frac{\partial^2 \bar{t}^\alpha}{\partial \bar{t}^\gamma \partial t^\beta} \right)_0 = \frac{d^2 \tau_s^\alpha}{dt^\gamma dt^\beta}, \qquad \left( \frac{\partial}{\partial \varepsilon^s} \frac{\partial^2 \bar{x}^k}{\partial t^\gamma \partial t^\beta} \right)_0 = \frac{d^2 \xi_s^k}{dt^\gamma dt^\beta},$$

which can be obtained from the defining transformations (6.20). Now, substitution of (6.27) and (6.26) into (6.24) yields, after renaming the indices, the fundamental invariance identities (6.23). Thus the proof is completed. □

The Noether theorem for the second-order, multiple integral problem follows directly from the invariance identities (6.23). By Exercise 6.5 we note that an equivalent form for (6.23) is

$$\frac{d}{dt^\alpha} (L\tau_s^\alpha) + \frac{\partial L}{\partial x^k} C_s^k + \frac{\partial L}{\partial \dot{x}_\alpha^k} \frac{d}{dt^\alpha} C_s^k + \frac{\partial L}{\partial \ddot{x}_{\alpha\beta}^k} \frac{d^2}{dt^\alpha dt^\beta} C_s^k = 0, \tag{6.28}$$

where

$$C_s^k \equiv \xi_s^k - \dot{x}_\gamma^k \tau_s^\gamma. \tag{6.29}$$

We obviously have

$$
\frac{\partial L}{\partial \dot{x}_\alpha^{\,k}} \frac{d}{dt^\alpha} C_s^{\,k} = \frac{d}{dt^\alpha}\left(\frac{\partial L}{\partial \dot{x}_\alpha^{\,k}} C_s^{\,k}\right) - C_s^{\,k} \frac{d}{dt^\alpha} \frac{\partial L}{\partial \dot{x}_\alpha^{\,k}}
$$

and

$$
\frac{\partial L}{\partial \ddot{x}_{\alpha\beta}^{\,k}} \frac{d^2}{dt^\alpha\,dt^\beta} C_s^{\,k} = \frac{d}{dt^\alpha}\left(\frac{\partial L}{\partial \ddot{x}_{\alpha\beta}^{\,k}} \frac{d}{dt^\beta} C_s^{\,k}\right) - \frac{d}{dt^\beta} \frac{\partial L}{\partial \ddot{x}_{\beta\alpha}^{\,k}} C_s^{\,k} + \frac{d}{dt^\alpha\,dt^\beta} \frac{\partial L}{\partial \ddot{x}_{\alpha\beta}^{\,k}} C_s^{\,k}.
$$

Substituting these expressions into (6.28) we get, for $s = 1, \ldots, r$,

$$
-E_k^{(2)} C_s^{\,k} = \frac{d}{dt^\alpha}\left[ L\tau_s^{\,\alpha} + \frac{\partial L}{\partial \dot{x}_\alpha^{\,k}} C_s^{\,k} + \frac{\partial L}{\partial \ddot{x}_{\alpha\beta}^{\,k}} \frac{dC_s^{\,k}}{dt^\beta} - \frac{d}{dt^\beta} \frac{\partial L}{\partial \ddot{x}_{\beta\alpha}^{\,k}} C_s^{\,k} \right], \quad (6.30)
$$

where $E_k^{(2)}$ are the second-order Euler expressions defined in Eq. (6.6). The $r$ identities (6.30) are the *Noether identities* for the second-order multiple integral problem. They clearly reduce to the Noether identities for the single integral case, Eqs. (6.16), when $m = 1$. From (6.30) we immediately obtain the following theorem giving conservation laws for the given problem.

**6.6  Theorem**  If the fundamental integral (6.2) is invariant under the $r$-parameter family of transformations (6.20), and if $x = x(t)$ is an extremal surface, then the following $r$ conservation laws hold true:

$$
\frac{d}{dt^\alpha}\left[ L\tau_s^{\,\alpha} + \left(\frac{\partial L}{\partial \dot{x}_\alpha^{\,k}} - \frac{d}{dt^\beta} \frac{\partial L}{\partial \ddot{x}_{\beta\alpha}^{\,k}}\right) C_s^{\,k} + \frac{\partial L}{\partial \ddot{x}_{\alpha\beta}^{\,k}} \frac{dC_s^{\,k}}{dt^\beta} \right] = 0 \quad (6.31)
$$

for $s = 1, \ldots, r$, where $C_s^{\,k}$ is defined by (6.29).

## 6.4  THE KORTEWEG–DEVRIES EQUATION

One of the fundamental equations of mathematical physics which arises in a number of physical situations is the so-called Korteweg–deVries equation (KdV equation for short). It is a third-order, nonlinear partial differential equation of the form

$$
u_t - 6uu_{xx} + u_{xxx} = 0, \quad (6.32)
$$

where subscripts denote partial differentiation with respect to spatial and time variables $x$ and $t$, respectively. The equation was first written down by Korteweg and deVries in 1895 in connection with the evolution of long water waves down canals of rectangular cross section. The equation also arises in plasma physics, in the study of an harmonic lattices, and in the propagation of waves in elastic rods. A survey of some of the results and applications of the KdV equation can be found in a recent article by Miura

[1]; an extensive bibliography is also contained in this paper. See also Whitham [1].

In this section our goal is to show how conservation laws for systems governed by the KdV equation can easily be derived from the Noether theorem. These calculations will provide a significant application of the results in the last section.

As the equation stands in (6.32), it is not possible to write down a Lagrange function which gives the KdV equation as the associated Euler–Lagrange equation. However, if we introduce the "potential" function

$$u = v_x, \tag{6.33}$$

then the KdV Eq. (6.32) becomes

$$v_{tx} - 6v_x v_{xx} + v_{xxxx} = 0, \tag{6.34}$$

which is a fourth-order partial differential equation. It is not difficult to see that Eq. (6.34) can be obtained from a variational principle with Lagrangian density given by

$$L = \tfrac{1}{2} v_t v_x - v_x^{\ 3} - \tfrac{1}{2} v_{xx}^2. \tag{6.35}$$

Note, therefore, that we may derive the KdV equation from a second-order variation problem.

From the form (6.35) of the Lagrangian density, it is immediately evident that the action integral

$$J(v) = \int L(t, x, v, v_t, v_x, v_{tt}, v_{xt}, v_{tx}, v_{xx})\, dt\, dx$$

is absolutely invariant under the transformations

$$\bar{t} = t, \qquad \bar{x} = x, \qquad \bar{v} = v + \varepsilon, \tag{6.36}$$

and

$$\bar{t} = t, \qquad \bar{x} = x + \varepsilon, \qquad \bar{v} = v, \tag{6.37}$$

and

$$\bar{t} = t + \varepsilon, \qquad \bar{x} = x, \qquad \bar{v} = v. \tag{6.38}$$

These three transformations clearly represent infinitesimal translations of the variables $v$, $t$, and $x$. Equations (6.36) through (6.38) are obviously equivalent to the three-parameter family of transformations

$$\bar{t} = t + \varepsilon^1, \qquad \bar{x} = x + \varepsilon^2, \qquad \bar{v} = v + \varepsilon^3, \tag{6.39}$$

where $\varepsilon^1$, $\varepsilon^2$, and $\varepsilon^3$ are the three independent parameters.

According to Theorem 6.6 the three resulting conservation laws take the form

$$\frac{d}{dt^\alpha}\left[ L\tau_s{}^\alpha + \left( \frac{\partial L}{\partial v_\alpha} - \frac{d}{dt^\beta}\frac{\partial L}{\partial v_{\alpha\beta}} \right)(\xi_s - v_\gamma\tau_s{}^\gamma) + \frac{\partial L}{\partial v_{\alpha\beta}}\frac{d}{dt^\beta}(\xi_s - v_\gamma\tau_s{}^\gamma) \right] = 0,$$

$$(6.40)$$

for $s = 1, 2, 3$. Here, $t^0 = t$, $t^1 = x$, and the subscripts $\alpha$, $\beta$, and $\gamma$ take the values 0 or 1, corresponding to $t$ and $x$, respectively. The infinitesimal generators are

$$\xi_s = \left.\frac{\partial \bar{v}}{\partial \varepsilon^s}\right|_0, \qquad \tau_s{}^0 = \left.\frac{\partial \bar{t}}{\partial \varepsilon^s}\right|_0, \qquad \tau_s{}^1 = \left.\frac{\partial \bar{x}}{\partial \varepsilon^s}\right|_0.$$

From the definition of the Lagrangian density, Eq. (6.35), it follows that

$$\frac{\partial L}{\partial v_t} = \tfrac{1}{2}v_x, \qquad \frac{\partial L}{\partial v_x} = \tfrac{1}{2}v_t - 3v_x{}^2, \qquad \frac{\partial L}{\partial v_{xx}} = -v_{xx}. \qquad (6.41)$$

All the other partial derivatives of $L$ vanish.

When $s = 3$, we have $\tau_3{}^0 = \tau_3{}^1 = 0$ and $\xi_3 = 1$. Hence (6.40) in this case becomes

$$\frac{\partial}{\partial t}\left(\frac{v_x}{2}\right) + \frac{\partial}{\partial x}\left(\frac{v_t}{2} - 3v_x{}^2 + \frac{\partial}{\partial x}v_{xx}\right) = 0$$

or

$$\tfrac{1}{2}v_{xt} + \tfrac{1}{2}v_{tx} + \frac{\partial}{\partial x}(-3v_x{}^2 + v_{xxx}) = 0.$$

Using (6.33) we can write this conservation law in the form

$$\frac{\partial u}{\partial t} + \frac{\partial}{\partial x}(-3u^2 + u_{xx}) = 0. \qquad (6.42)$$

When $s = 2$, we have $\tau_2{}^0 = \xi_2 = 0$ and $\tau_2{}^1 = 1$. In this case (6.40) becomes

$$\frac{\partial}{\partial t}(-\tfrac{1}{2}v_x{}^2) + \frac{\partial}{\partial x}(2v_x{}^3 + \tfrac{1}{2}v_{xx}^2 - v_xv_{xxx}) = 0.$$

Using (6.33), this equation gives the conservation law,

$$\frac{\partial u^2}{\partial t} + \frac{\partial}{\partial x}(-4u^3 - u_x{}^2 + 2uu_{xx}) = 0. \qquad (6.43)$$

In the case $s = 1$ the resulting conservation law is

$$\frac{\partial}{\partial t}(u^3 + \tfrac{1}{2}u_x{}^2) + \frac{\partial}{\partial x}(-\tfrac{9}{2}u^4 + 3u^2 u_{xx} - 6uu_x{}^2 + u_x u_{xxx} - \tfrac{1}{2}u_{xx}^2) = 0. \quad (6.44)$$

We leave this calculation as an exercise. For some physical systems, the three conservation laws (6.42), (6.43), and (6.44) can be interpreted to give conservation of mass, momentum, and energy, respectively.

As it turns out, the KdV equation is also Galilean invariant. That is, the KdV equation is left unchanged by the transformation

$$\bar{t} = t, \qquad \bar{x} = x - \varepsilon t, \qquad \bar{u} = u + \frac{\varepsilon}{6}. \quad (6.45)$$

An application of Noether's theorem yields the conservation law

$$\frac{\partial}{\partial t}(xu + 3tu^2) + \frac{\partial}{\partial x}(-3u^2 x + xu_{xx} - u_x - 12tu^3 + 6tuu_{xx} - 3tu_x{}^2) = 0.$$

$$(6.46)$$

Again, we leave this straightforward calculation to the reader.

## 6.5  BIBLIOGRAPHIC NOTES

Higher-order variation problems are discussed in most textbooks on the calculus of variations. An advanced-level monograph with an extensive bibliography has been written by Grässer [1].

Invariant higher-order problems were first considered by Noether [1] in her original paper. Anderson [1] puts Noether's work in more modern form. Barut [1] can also be consulted.

The fundamental invariance identifies for second-order problems are discussed in Logan and Blakeslee [1], and Blakeslee and Logan [1, 2]. These papers cover single and multiple integral problems with applications to scalar fields and covariant vector fields.

The $p$th-order invariant single integral problem is discussed by Blakeslee [1]. We shall document some of his results. Consider the variation problem defined by

$$J(x) = \int_a^b L(t, x, D^1 x, \ldots, D^p(x))\, dt \to \min,$$

where $D^k = d^k/dt^k$ and $x = (x^1, \ldots, x^n)$. Then we have:

**6.7   Theorem**   If $J(x)$ is absolutely invariant under the $r$-parameter family of transformations

$$\bar{t} = t + \tau_s(t, x)\varepsilon^s + o(\varepsilon), \qquad \bar{x}^k = x^k + \xi_s{}^k(t, x)\varepsilon^s + o(\varepsilon),$$

then the Lagrangian and its derivatives satisfy the $r$ identities

$$L\frac{d\tau_s}{dt} + \frac{\partial L}{\partial t}\tau_s + \sum_{k=1}^{n}\frac{\partial L}{\partial x^k}\xi_s{}^k$$

$$+ \sum_{k=1}^{n}\sum_{i=1}^{p}\left(\frac{\partial L}{\partial(D^i x^k)}D^i(\xi_s{}^k - \dot{x}^k\tau_s) + D^{i+1}x^k\tau_s\right) = 0.$$

From this we get:

**6.8   Theorem**   If, in addition to the hypotheses of Theorem 6.7, the Euler–Lagrange equations

$$\frac{\partial L}{\partial x^k} - D^1\frac{\partial L}{\partial \dot{x}^k} + \cdots + (-1)^p D^p\frac{\partial L}{\partial(D^p x^k)} = 0$$

are satisfied, then the following $r$ conservation laws hold true:

$$L\tau_s + \sum_{k=1}^{n}\sum_{j=1}^{p}P_{k,j}D^{j-1}(\xi_s{}^k - \dot{x}^k\tau_s) = \text{constant},$$

where

$$P_{k,j} \equiv \sum_{i=0}^{p-j}(-1)^i D^i \frac{\partial L}{\partial(D^{i+j}x^k)}.$$

## EXERCISES

**6-1**   Let

$$J = \int_{t_0}^{t_1}\int_0^l (\tfrac{1}{2}\rho u_t{}^2 - \tfrac{1}{2}k u_{xx}^2)\, dx\, dt.$$

Show that $J$ is invariant under translations of $t$, $x$, and $u$ and derive the resulting conservation laws.

**6-2**   Find the Euler–Lagrange equation associated with the Lagrangian

$$L = -\sum_{\alpha=1}^{n} x \frac{\partial^2 x}{\partial t^\alpha\, \partial t^\alpha}.$$

What are the conservation laws?

**6-3**  Find an extremal of the functional

$$J(x) = \int_0^1 (1 + \ddot{x}^2)\, dt$$

subject to the conditions

$$x(0) = 0, \quad x(1) = 0, \quad \dot{x}(0) = 1, \quad \dot{x}(1) = 1.$$

What are some constants of the motion?

**6-4**  Write the Euler–Lagrange equation for the functional

$$J(u) = \int (-(u_{xx} + u_{yy})^2 + 2(1 - \mu)(u_{xx}u_{yy} - u_{xy}^2))\, dt\, dx\, dy,$$

where $\mu$ is a constant. What are the conservation laws?

**6-5**  Show that identities (6.13) and (6.23) may be written, respectively, as

$$\frac{d}{dt}(L\tau_s) + \frac{\partial L}{\partial x^k}(\xi_s{}^k - \dot{x}^k\tau_s) + \frac{\partial L}{\partial \dot{x}^k}\frac{d}{dt}(\xi_s{}^k - \dot{x}^k\tau_s) + \frac{\partial L}{\partial \ddot{x}^k}\frac{d^2}{dt^2}(\xi_s{}^k - \dot{x}^k\tau_s) = 0$$

and

$$\frac{d}{dt^\alpha}(L\tau_s{}^\alpha) + \frac{\partial L}{\partial x^k}(\xi_s{}^k - \dot{x}_\gamma{}^k\tau_s{}^\gamma) + \frac{\partial L}{\partial \dot{x}_\alpha{}^k}\frac{d}{dt^\alpha}(\xi_s{}^k - \dot{x}_\gamma{}^k\tau_s{}^\gamma)$$

$$+ \frac{\partial L}{\partial \ddot{x}_{\alpha\beta}^k}\frac{d^2}{dt^\alpha\, dt^\beta}(\xi_s{}^k - \dot{x}_\gamma{}^k\tau_s{}^\gamma) = 0.$$

**6-6**  From first principles, derive the Euler–Lagrange equations for the single integral problem

$$\int_a^b L(t, x, \dot{x}, \ddot{x})\, dt \to \min, \qquad x \in C^4[a, b].$$

**6-7**  The equation

$$u_t + au_x + bu_{xxx} = 0,$$

where $a$ and $b$ are constants is the linearized form of the Korteweg–deVries equation which occurs in shallow water wave theory. By introducing a potential function $u = v_x$, determine a Lagrangian $L = L(t, x, v, v_t, v_x, v_{tx}, v_{tt}, v_{xx})$ which yields the linearized Korteweg–deVries equation as the associated Euler–Lagrange equation, and find the conservation laws associated with translational invariance.

**6-8**  The equation

$$u_{tt} - a\nabla^2 u - b\nabla^2 u_{tt} = 0, \qquad a > 0, \quad b > 0,$$

appears in elasticity for longitudinal waves in bars, in water waves in the Boussinesq approximation for long waves, and in plasma waves. Determine a variational principle from which this equation follows. Find conservation laws.

**6-9**  Use Noether's theorem to derive conservation laws for systems governed by the *modified* KdV equation

$$u_t - 6u^2 u_x + u_{xxx} = 0.$$

**6-10**  By substituting an assumed solution in the KdV equation of the form of a traveling wave, i.e., $u(t, x) = U(x - ct)$, show that $U$ satisfies the equation

$$U''' - (6U + c)U' = 0.$$

Hence, show that

$$U'^2 - 2U^3 - cU^2 - AU = B,$$

where $A$ and $B$ are constants of integration. Integrate this equation to get

$$\int \frac{dU}{\sqrt{B + AU + cU^2 + 2U^3}} + D = \pm(x - ct),$$

where $D$ is a constant. Under the assumption that $\lim_{|x| \to \infty} U = 0$, show that

$$u(t, x) = -\tfrac{1}{2}a^2 \, \text{sech}\left(\frac{a}{2}(x - a^2 t)\right),$$

when the center of the symmetrical wave at $t = 0$ is at $x = 0$.

**6-11**  Derive the conservation law (6.44) from Noether's theorem.

**6-12**  Derive the conservation law (6.45) which follows from the invariance of the KdV equation under (6.45).

# CHAPTER 7

# Conformally Invariant Problems

## 7.1  CONFORMAL TRANSFORMATIONS

Conformal transformations have enjoyed considerable attention in problems of mathematics for a long time, especially in theoretical and practical applications to the Dirichlet problem. In the first decade of this century these transformations gained even more relevance to physical problems when Bateman [1] and Cunningham [1] showed that Maxwell's equations in electromagnetic theory were not only invariant under the ten-parameter Poincaré group consisting of space–time translations and the proper, homogeneous Lorentz transformations, but also under the larger special conformal group. The special conformal transformations form a fifteen-parameter local Lie group which, in addition to translations and Lorentz transformations, include a space–time dilation and space–time inversions.

The aim of the present chapter is to investigate variation problems which are invariant under these special conformal transformations. The original work in this area was done by Bessel-Hagen [1] in 1921 who showed that the variation integral for electrodynamics *in vacuo* was invariant under the special conformal transformations and he derived the resulting fifteen conservation laws from Noether's theorem. Besides discussing conservation laws for both scalar and vector fields in this chapter, we shall also observe that the fundamental invariance identities lead to some interesting results. For, as it turns out, we shall be able in some cases to characterize those Lagrangians whose corresponding action integral is conformally invariant.

However, before beginning an investigation of variation problems, we shall introduce some basic concepts in the study of conformal transformations. Rather than go deeply into the theory and applications, we shall be satisfied to state some fundamental definitions and results which will be useful in our analysis.

Conformal transformations can be formulated in several different ways. Classically, we think of a conformal transformation on some space $X_n$ as a point transformation on $X_n$ which preserves angles and which preserves distances up to some scalar multiple. This implies, of course, that the space is equipped with a metric tensor $g_{\alpha\beta}$ which allows distances and angles to be defined. Recalling the comments in Chapter 5, the length of an arc element from $t = (t^\alpha)$ to $t + dt = (t^\alpha + dt^\alpha)$ is given by

$$ds^2(t) = g_{\alpha\beta}(t)\, dt^\alpha\, dt^\beta. \tag{7.1}$$

The angle $\theta$ between two vectors $a^\alpha$ and $b^\alpha$ at $t$ is defined, generalizing the elementary definition $\cos\theta = (a \cdot b)/|a||b|$, by

$$\cos\theta = \frac{g_{\alpha\beta}(t)a^\alpha b^\beta}{\sqrt{g_{\alpha\beta}(t)a^\alpha a^\beta}\sqrt{g_{\alpha\beta}(t)b^\alpha b^\beta}}. \tag{7.2}$$

It is easy to see that both $ds^2(t)$ and $\cos\theta$ are scalars or invariants; i.e., they do not depend on the particular coordinate system chosen for their representation. Now, from these ideas we can formulate a definition of a conformal transformation. We shall denote different points of $X_n$ by $t = (t^\alpha)$, $y = (y^\alpha)$, $z = (z^\alpha), \ldots$, with $\bar{t} = (\bar{t}^\alpha)$, $\bar{y} = (\bar{y}^\alpha)$, $\bar{z} = (\bar{z}^\alpha), \ldots$, denoting those same points in different coordinate systems. Then, by a conformal point transformation we mean a transformation $t \to y$ for which the line element $ds^2(t)$ at $t$ is related to the line element $ds^2(y)$ at $y$ by a positive scalar function. More precisely, we have:

**7.1  Definition**   A one-to-one mapping

$$y^\alpha = f^\alpha(t)$$

on $X_n$ is called a *conformal point transformation* if

$$g_{\alpha\beta}(y)\frac{\partial y^\alpha}{\partial t^\mu}\frac{\partial y^\beta}{\partial t^\nu} = \sigma(t)g_{\mu\nu}(t) \tag{7.3}$$

for all $t \in X_n$, for some scalar function $\sigma(t) > 0$.

It will be shown in an exercise that (7.3) is equivalent to the condition $ds^2(y) = \sigma(t) \, ds^2(t)$. Moreover, it is not difficult to observe that angles are preserved under the conditions of Definition 7.1. The function $\sigma(t)$ is known as the *conformal factor* of the transformation.

An alternate formulation can be given in terms of a redefinition of the metric tensor.

**7.2  Definition**  If

$$\hat{g}_{\alpha\beta}(t) \equiv \sigma(t) g_{\alpha\beta}(t) \qquad (t \in X_n), \tag{7.4}$$

where $\sigma(t) > 0$, then $\hat{g}_{\alpha\beta}$ is called a *conformal transformation of the metric tensor*.

Clearly, if a conformal point transformation is given, then

$$\hat{g}_{\alpha\beta}(t) = g_{\mu\nu}(f(t)) \frac{\partial y^\mu}{\partial t^\alpha} \frac{\partial y^\nu}{\partial t^\beta}$$

defines a conformal transformation of the metric tensor. Conversely, to a conformal transformation $\hat{g}_{\alpha\beta}$ of the metric tensor corresponds a conformal point transformation through the solution of the system of partial differential equations

$$g_{\mu\nu}(y) \frac{\partial y^\mu}{\partial t^\alpha} \frac{\partial y^\nu}{\partial t^\beta} = \sigma(t) g_{\alpha\beta}(t)$$

for $y^\alpha = y^\alpha(t) \equiv f^\alpha(t)$. Thus we see that a conformal transformation of the metric tensor is equivalent to a conformal point transformation.

Up to now our study of invariant variation problems has involved transformations which are changes of coordinate systems, or coordinate transformations, rather than the point transformations defined above. Of the two concepts, a point transformation (sometimes called an active transformation) would seem to be more fundamental than a coordinate transformation (passive transformation), the latter being only a relabeling of points. However, from the viewpoint of physics, we can regard coordinate transformations as a relation between two observers viewing the same domain of points; i.e., we identify a coordinate system with an observer. Fortunately, there is a natural and simple connection between these two types of transformations. Suppose we are given a point transformation on $X_n$,

$$y^\alpha = f^\alpha(t).$$

Since the $f^\alpha$ are one-to-one, we may solve these equations for the $t^\alpha$ to obtain

$$t^\alpha = F^\alpha(y).$$

Using these same functions $F^\alpha$, we can define a coordinate transformation

$$\bar{t}^\alpha = F^\alpha(t).$$

Also, by arguing in the opposite direction we can obviously show that a unique point transformation results from a given coordinate transformation and therefore these two notions are equivalent. From these definitions, we note that in the barred coordinate system the point $y$ has coordinates

$$\bar{y}^\alpha = F^\alpha(y) = F^\alpha(f(t)) \doteq t^\alpha,$$

where the dot indicates equality only in the coordinate systems indicated in the equation. Thus, the correlation between point and coordinate transformations amounts to identifying points in such a way that the new coordinates of $y$ are the same, numerically, as the old or unbarred coordinates of $t$, where $y^\alpha = f^\alpha(t)$.

**7.1　Example**　Consider the point transformation from $R^1$ to $R^1$ defined by $y = f(t) \equiv 3t + 1$. Inverting, we obtain $t = (y - 1)/3 \equiv F(y)$. Therefore, the corresponding coordinate transformation is $\bar{t} = (t - 1)/3$. Note that, for example, if $t = 5$, then $y = 16$; but in the new system, $\bar{y} = 5$. Thus, the identification is made such that $t \doteq \bar{y}$.

　　As a part of our development we shall now devote a few paragraphs to the study of conformal transformations on $R^n$, $n > 2$, where the fundamental quadratic form $g_{\alpha\beta} \, dt^\alpha \, dt^\beta$ which defines the length of an arc element is assumed to be *positive definite*, i.e.,

$$g_{\alpha\beta} \, dt^\alpha \, dt^\beta > 0 \qquad \text{for all} \quad dt = (dt^\alpha) \neq 0. \tag{7.5}$$

(Similarly, we can define *negative definite* by replacing $>$ with $<$. A metric or quadratic form which is not positive definite or negative definite is called *indefinite*.) As always, we assume that $g_{\alpha\beta}$ is symmetric, i.e., $g_{\alpha\beta} = g_{\beta\alpha}$. As we shall observe later, the basic conformal transformations on $R^n$ are the translations, rotations, dilations, and inversions. In turn, we shall now comment on these transformations.

　　A *translation* on $R^n$ is a point transformation of the form

$$y^\alpha = t^\alpha - \varepsilon^\alpha \qquad \text{(translation)}, \tag{7.6}$$

where $\varepsilon = (\varepsilon^\alpha)$ is a given vector in $R^n$. It is clear from our preceding remarks that (7.6) corresponds to the coordinate transformation

$$\bar{t}^\alpha = t^\alpha + \varepsilon^\alpha. \tag{7.7}$$

In this case it is easy to see that (7.6) is conformal (it is geometrically obvious since translations preserve angles) with conformal factor $\sigma(t) = 1$. For, using (7.3),

$$g_{\alpha\beta} \frac{\partial y^\alpha}{\partial t^\mu} \frac{\partial y^\beta}{\partial t^\nu} = g_{\alpha\beta} \delta_\mu{}^\alpha \delta_\nu{}^\beta = g_{\mu\nu}.$$

A *rotation* in $R^n$ about the origin is a point transformation which can be represented by an orthogonal matrix $A = (a_\beta{}^\alpha)$. That is,

$$y^\alpha = a_\beta{}^\alpha t^\beta \qquad \text{(rotation)}, \tag{7.8}$$

where $A^T = A^{-1}$. Then the corresponding coordinate transformation is given by

$$\bar{t}^\alpha = b_\beta{}^\alpha t^\beta,$$

where $A^{-1} = (b_\beta{}^\alpha)$. Rotations clearly preserve angles, and their conformal factor is $\sigma(t) = 1$. This follows from

$$g_{\alpha\beta} \frac{\partial y^\alpha}{\partial t^\mu} \frac{\partial y^\beta}{\partial t^\nu} = g_{\alpha\beta} a_\gamma{}^\alpha \delta_\mu{}^\gamma a_\lambda{}^\beta \delta_\nu{}^\lambda = g_{\mu\nu}.$$

The rotations and translations are collectively referred to as *motions*.

A *dilation*, or magnification, is a mapping of the form

$$y^\alpha = \gamma t^\alpha \qquad (\gamma > 0) \qquad \text{(dilation)}. \tag{7.9}$$

By inverting this transformation we are able to write the corresponding coordinate transformation

$$\bar{t}^\alpha = \frac{1}{\gamma} t^\alpha.$$

The conformal factor in this case can again be calculated using (7.3):

$$g_{\alpha\beta} \frac{\partial y^\alpha}{\partial t^\mu} \frac{\partial y^\beta}{\partial t^\nu} = g_{\alpha\beta} \gamma^2 \delta_\mu{}^\alpha \delta_\nu{}^\beta = \gamma^2 g_{\mu\nu}.$$

Thus $\sigma(t) = \gamma^2$.

*Inversions* with respect to a unit sphere centered at the origin are point transformations of the form

$$y^\alpha = \frac{t^\alpha}{g_{\mu\nu} t^\mu t^\nu} \qquad (t \neq 0) \qquad \text{(inversion)}. \tag{7.10}$$

Such a transformation has the property that both $t$ and its image $y$ lie on the same line through the origin 0 and $|\overline{0y}||\overline{0t}| = 1$, i.e., $\sqrt{g_{\alpha\beta} y^\alpha y^\beta}\sqrt{g_{\mu\nu} t^\mu t^\nu} = 1$. In order to invert (7.10), we note that an inversion followed by another inversion gives back the original point. In more precise terms, if $y = f(t)$ is an inversion, then $t = f(y) = f(f(t))$, and so

$$f \circ f = I, \tag{7.11}$$

where $I$ is the identity transformation, and where the circle $\circ$ denotes composition. Any transformation $f$ satisfying (7.11) is called an *involution*; if $f$ is invertible, then it follows from (7.11) that

$$f = f^{-1},$$

where $f^{-1}$ denotes the inverse of $f$.

Therefore, since inversions are involutions, we conclude from (7.10) that

$$t^\alpha = \frac{y^\alpha}{g_{\mu\nu} y^\mu y^\nu}.$$

As a result, we obtain the coordinate transformation

$$\bar{t}^\alpha = \frac{t^\alpha}{g_{\mu\nu} t^\mu t^\nu}$$

associated with the point transformation (7.10). The determination of the conformal factor $\sigma(t)$ is straightforward; we leave it as an exercise to show that

$$\sigma(t) = (g_{\mu\nu} t^\mu t^\nu)^{-2}. \tag{7.12}$$

We now sum up our remarks on conformal point transformations on $R^n$ for a positive definite metric by stating a well-known theorem which characterizes all conformal transformations. For $n = 3$ the theorem is due to Liouville, and it is known as Liouville's theorem.

**7.1  Theorem**  Every real conformal point transformation in an $n$-dimensional ($n > 2$) Euclidean space ($R^n$) with a positive definite metric $g_{\alpha\beta}$ can be composed of a motion and an inversion or a motion and a dilation.  □

Liouville's theorem does not hold in $R^n$ when the metric is indefinite, which brings us to the next topic, namely, the form of the conformal transformations on $R^4$ when the distance is defined by

$$ds^2 = -c^2 t^2 + x^2 + y^2 + z^2.$$

To state a theorem analogous to Liouville's theorem for indefinite metrics, we need to introduce a new class of transformations which are products of

inversions and translations. These transformations are point transformations
of the form

$$y^\alpha = \frac{t^\alpha - \varepsilon^\alpha t_\lambda t^\lambda}{1 - 2\varepsilon_\beta t^\beta + \varepsilon_\beta \varepsilon^\beta t_\lambda t^\lambda},$$  (7.13)

where we have used the tensorial notation $t_\lambda t^\lambda = g_{\mu\lambda} t^\mu t^\lambda$.

Geometrically, we can realize (7.13) by performing, in succession, an
inversion $f_1$ through a unit circle centered at the origin,

$$f_1: \quad z^\alpha = \frac{t^\alpha}{t_\lambda t^\lambda},$$  (7.14)

an inversion $f_2$ through a unit circle centered at $\varepsilon = (\varepsilon^\alpha)$,

$$f_2: \quad w^\alpha = \frac{z^\alpha - \varepsilon^\alpha}{(z_\lambda - \varepsilon_\lambda)(z^\lambda - \varepsilon^\lambda)} + \varepsilon^\alpha,$$  (7.15)

and finally a translation $f_3$ by the vector $\varepsilon = (\varepsilon^\alpha)$,

$$f_3: \quad y^\alpha = w^\alpha - \varepsilon^\alpha.$$  (7.16)

In fact, Eq. (7.13) follows by substituting (7.15) into (7.16) and then substituting (7.14) into that result. Figure 8 illustrates these mappings; the
transformation (7.13) is the composition $f_3 \circ f_2 \circ f_1$.

In order to find the coordinate transformation associated with (7.13)
we must invert the mapping $f_3 \circ f_2 \circ f_1$. Since it is not evident how to
accomplish this by solving for $t^\alpha$ in terms of the $y^\alpha$ in (7.13), we proceed

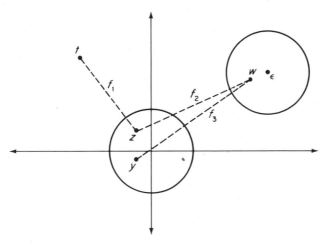

**FIGURE 8**

indirectly by noting that $(f_3 \circ f_2 \circ f_1)^{-1} = f_1^{-1} \circ f_2^{-1} \circ f_3^{-1}$. Since $f_1$ and $f_2$ are involutions, it follows from (7.14) and (7.15) that

$$f_1^{-1}: \quad t^\alpha = \frac{z^\alpha}{z_\lambda z^\lambda} \tag{7.17}$$

and

$$f_2^{-1}: \quad z^\alpha = \frac{w^\alpha - \varepsilon^\alpha}{(w_\lambda - \varepsilon_\lambda)(w^\lambda - \varepsilon^\lambda)} + \varepsilon^\lambda. \tag{7.18}$$

Equation (7.16) easily gives

$$f_3^{-1}: \quad w^\alpha = y^\alpha + \varepsilon^\alpha. \tag{7.19}$$

To compute $f_1^{-1} \circ f_2^{-1} \circ f_3^{-1}$ we substitute (7.19) into (7.18) and that result into (7.17) to obtain

$$t^\alpha = \frac{y^\alpha + \varepsilon^\alpha y_\lambda y^\lambda}{1 + 2\varepsilon_\lambda y^\lambda + \varepsilon_\mu \varepsilon^\mu y_\lambda y^\lambda},$$

which is the inverse of (7.13). Consequently, the corresponding coordinate transformation is given by

$$\bar{t}^\alpha = \frac{t^\alpha + \varepsilon^\alpha t_\lambda t^\lambda}{1 + 2\varepsilon_\lambda t^\lambda + \varepsilon_\mu \varepsilon^\mu t^\lambda t_\lambda}. \tag{7.20}$$

We are now in position to state a generalization of Liouville's theorem to $R^n$ with an indefinite metric. The theorem is due to Haantjes [1].

**7.2  Theorem**  Every real conformal point transformation in an $n$-dimensional $(n > 2)$ Euclidean space $(R^n)$ with indefinite metric $g_{\alpha\beta}$ is composed of a motion and a dilation, a motion and an inversion, or a motion and a transformation of the form (7.13).  □

The proof given by Haantjes is based on the fact that the conformal transformations above can be characterized as those transformations which map a flat space to a flat space. A different proof, modeled after Liouville's work, has been given by Hsu [1].

The translations (7.6), the rotations (7.8), the dilations (7.9), and the transformations (7.13) form the set of so-called *special conformal transformations*. We are interested, for the purpose of studying invariant variation problems, in the infinitesimal representations of those transformations in Minkowski space, or at least a representation in terms of finitely many parameters. Fortunately, we have such a representation already; letting

$$\varepsilon^0, \ldots, \varepsilon^3, \varepsilon_{01}, \varepsilon_{02}, \varepsilon_{03}, \varepsilon_{12}, \varepsilon_{13}, \varepsilon_{23}, \gamma, \eta^0, \eta^1, \eta^2, \text{ and } \eta^4$$

denote fifteen independent parameters, it is clear that we can write down the special conformal transformations as

(i)   $\bar{t}^\alpha = t^\alpha + \varepsilon^\alpha$                                                                    (translations),

(ii)   $\bar{t}^\alpha = t^\alpha + \sum_{\beta=0}^{3} g^{\beta\beta}\varepsilon_{\alpha\beta}t^\beta,$      $\varepsilon_{\alpha\beta} = -\varepsilon_{\beta\alpha}$      (rotations),

$$(7.21)$$

(iii)   $\bar{t}^\alpha = t^\alpha + \gamma t^\alpha$                                                                    (dilation),

(iv)   $\bar{t}^\alpha = (t^\alpha + \eta^\alpha t_\lambda t^\lambda)(1 + 2\eta_\lambda t^\lambda + \eta_\lambda \eta^\lambda t_\beta t^\beta)^{-1}$      (inversions),

where $g_{\alpha\beta} = -1$ for $\alpha = \beta = 0$, $g_{\alpha\beta} = 1$ for $\alpha = \beta = 1, 2, 3$, and $g_{\alpha\beta} = 0$ for $\alpha \neq \beta$. Here we have chosen the coordinate transformations which is consistent with our earlier formulation. For brevity, we shall refer to the transformations (7.21iv) as inversions, realizing however that they are actually products of inversions and translations. In much of the physics literature, these transformations are referred to as acceleration transformations and indeed they can be shown as such. We recognize the rotations (7.21ii) as the proper, homogeneous Lorentz transformations derived in Section 5.4. We recall that the Lorentz transformations were defined as those linear transformations which left the quadratic form $g_{\alpha\beta} \, dt^\alpha \, dt^\beta$ invariant. Thus, Lorentz transformations are conformal transformations with conformal factor unity, and they can be regarded as rotations of four-dimensional Minkowski space. The paper of Fulton, Rohrlich, and Witten [1] can be consulted for a thorough discussion of the role of conformal transformations in physical problems as well as for additional bibliographical information.

## 7.2   CONFORMAL INVARIANCE IDENTITIES FOR SCALAR FIELDS

In this section we shall examine the consequences of imposing conformal invariance upon a multiple integral variation problem over a region in four-dimensional space–time whose Lagrangian depends upon a scalar function and its first derivatives. Using the fundamental invariance identities, it is possible to go a long way toward characterizing such Lagrangians.

The formulation of the variation problem is the same as that which occurs in Section 5.6. Thus, let $x(t) = x(t^0, \ldots, t^3)$ be a scalar function defined on a resion $D$ of space–time; i.e., $\bar{x}(\bar{t}) = x(t)$. We then consider the variation integral

$$J(x) = \int_D L(t, x, \dot{x}_1, \dot{x}_2, \dot{x}_3, \dot{x}_4) \, dt^0 \cdots dt^3, \qquad (7.22)$$

where $\dot{x}_\alpha \equiv \partial x/\partial t^\alpha$, $\alpha = 0, \ldots, 3$.

In order to write down the invariance identities, it is necessary to determine the infinitesimal generators of the special conformal transformations (7.21). We record this result in the following lemma.

**7.1  Lemma**  The infinitesimal generators of the special conformal transformations (7.21i–iv) are

(i)   $\tau_\lambda{}^\alpha = \delta_\lambda{}^\alpha$                    (translations),

(ii)  $\tau_{\lambda\mu}^\alpha = g^{\mu\mu}\delta_\lambda{}^\alpha t^\mu - g^{\lambda\lambda}\delta_\mu{}^\alpha t^\lambda$        (rotations),

(iii)     $\tau^\alpha = t^\alpha$                    (dilations),        (7.23)

(iv)  $\tau_\lambda{}^\alpha = t^\beta t_\beta \delta_\lambda{}^\alpha - 2t^\alpha t_\lambda$        (inversions),

where, in (ii), $(\lambda, \mu) \in S = \{(0, 1), (0, 2), (0, 3), (1, 2), (1, 3), (2, 3)\}$ and no summation is intended over $\lambda$ and $\mu$.

*Proof*   In the case of the translations (7.21i), clearly

$$\tau_\lambda{}^\alpha \equiv \left.\frac{\partial \bar{t}^\alpha}{\partial \varepsilon^\lambda}\right|_{\varepsilon=0} - \delta_\lambda{}^\alpha.$$

For the dilations (7.21iii),

$$\tau^\alpha \equiv \left.\frac{\partial \bar{t}^\alpha}{\alpha\gamma}\right|_{\gamma=0} = t^\alpha.$$

For inversions, a straightforward calculation using (7.21iv) yields

$$\tau_\lambda{}^\alpha \equiv \left.\frac{\partial \bar{t}^\alpha}{\partial \eta^\lambda}\right|_{\eta=0} = t^\beta t_\beta \delta_\lambda{}^\alpha - 2t^\alpha t_\lambda.$$

For rotations, from (7.21ii) it follows that

$$\tau_{\lambda\mu}^\alpha \equiv \left.\frac{\partial \bar{t}^\alpha}{\partial \varepsilon_{\lambda\mu}}\right|_{\varepsilon=0} = \left(\frac{\partial}{\partial \varepsilon_{\lambda\mu}}\sum_\beta g^{\beta\beta}\varepsilon_{\alpha\beta}t^\beta\right)_{\varepsilon=0}$$

$$= \sum_{\beta>\alpha} g^{\beta\beta}t^\beta \frac{\partial}{\partial \varepsilon_{\lambda\mu}}\varepsilon_{\alpha\beta} - \sum_{\beta<\alpha} g^{\beta\beta}t^\beta \frac{\partial}{\partial \varepsilon_{\lambda\mu}}\varepsilon_{\beta\alpha}$$

$$= \sum_\beta (g^{\beta\beta}t^\beta \delta_\lambda{}^\alpha \delta_\mu{}^\beta - g^{\beta\beta}t^\beta \delta_\lambda{}^\beta \delta_\mu{}^\alpha)$$

$$= g^{\mu\mu}t^\mu \delta_\lambda{}^\alpha - g^{\lambda\lambda}t^\lambda \delta_\mu{}^\alpha. \qquad \square$$

We can now state the following theorem.

**7.3 Theorem** If the fundamental integral (7.22) is absolutely invariant under the special conformal transformations (7.21), then the Lagrangian $L$ must satisfy the fifteen identities

(i) $\quad \dfrac{\partial L}{\partial t^{\alpha}} = 0, \qquad \alpha = 0, \ldots, 3;$

(ii) $\quad \dot{x}^{\alpha} \dfrac{\partial L}{\partial \dot{x}_{\alpha}} = 4L;$

$$(7.24)$$

(iii) $\quad \dot{x}_{\mu} \dfrac{\partial L}{\partial \dot{x}_{\lambda}} g^{\lambda\lambda} - \dot{x}_{\lambda} \dfrac{\partial L}{\partial \dot{x}_{\mu}} g^{\mu\mu} = 0, \qquad (\lambda, \mu) \in S \quad \text{(no sums)};$

(iv) $\quad t_{\alpha} \dot{x}_{\lambda} \dfrac{\partial L}{\partial \dot{x}_{\alpha}} - g_{\alpha\lambda} \dot{x}_{\beta} t^{\beta} \dfrac{\partial L}{\partial \dot{x}_{\alpha}} = 0, \qquad \lambda = 0, \ldots, 3.$

*Proof* Since the field $x = x(t)$ is a scalar field, it follows from Eq. (5.55) that the fundamental invariance identities take the form

$$\frac{\partial L}{\partial t^{\alpha}} \tau_{\lambda}{}^{\alpha} - \frac{\partial L}{\partial \dot{x}_{\alpha}} \frac{d\tau_{\lambda}{}^{\beta}}{dt^{\alpha}} \dot{x}_{\beta} + L \frac{d\tau_{\lambda}{}^{\alpha}}{dt^{\alpha}} = 0. \tag{7.25}$$

In Theorem 5.2 it was shown that (7.24i) and (7.24iii) are direct consequences of (7.25) when the transformations were given by the translations (7.21i) and rotations (7.21ii), respectively. Let us then consider dilations. From Lemma 7.1 we have

$$\tau^{\alpha} = t^{\alpha},$$

and therefore

$$\frac{d\tau^{\beta}}{dt^{\alpha}} = \delta_{\alpha}{}^{\beta}, \qquad \frac{d\tau^{\beta}}{dt^{\beta}} = 4.$$

Substitution of these quantities into (7.25) gives (7.24ii). Now we consider the inversions, where the infinitesimal generators are given by (7.23iv). In this case

$$\frac{d\tau_{\lambda}{}^{\alpha}}{dt^{\gamma}} = g_{\mu\beta} \delta_{\lambda}{}^{\alpha}(t^{\mu}\delta_{\gamma}{}^{\beta} + t^{\beta}\delta_{\gamma}{}^{\mu}) - 2g_{\mu\lambda}(t^{\mu}\delta_{\gamma}{}^{\alpha} + t^{\alpha}\delta_{\gamma}{}^{\mu})$$

$$= 2t_{\gamma}\delta_{\lambda}{}^{\alpha} - 2t_{\lambda}\delta_{\gamma}{}^{\alpha} - 2g_{\gamma\lambda}t^{\alpha}, \tag{7.26}$$

whence

$$\frac{d\tau_{\lambda}{}^{\alpha}}{dt^{\alpha}} = -8t_{\lambda}. \tag{7.27}$$

Substituting (7.26) and (7.27) into (7.25), we get

$$\dot{x}_\beta \frac{\partial L}{\partial \dot{x}_\alpha}(2t_\alpha\delta_\lambda{}^\beta - 2t_\lambda\delta_\alpha{}^\beta - 2g_{\alpha\lambda}t^\beta) + 8t_\lambda L = 0.$$

After rearranging and collecting terms, we obtain

$$t_\lambda\left(4L - \dot{x}_\alpha\frac{\partial L}{\partial \dot{x}_\alpha}\right) + t^\beta\left(g_{\alpha\beta}\dot{x}_\lambda\frac{\partial L}{\partial \dot{x}_\alpha} - g_{\alpha\lambda}\dot{x}_\beta\frac{\partial L}{\partial \dot{x}_\alpha}\right) = 0,$$

which, according to (7.24ii), reduces to (7.24iv).   $\square$

We emphasize that the conformal identities (7.24) are not conservation laws, but rather invariance identities which are necessary conditions for the fundamental integral (7.22) to be invariant under the special conformal transformations.

By integrating (7.24) we can discover the form of classes of Lagrangians for which the fundamental integral is conformally invariant. The key to such a calculation is to notice that the left-hand side of (7.24iii) is anti-symmetric in the indices $\lambda$ and $\mu$, and therefore (7.24iii) holds for all $\lambda, \mu = 0, \ldots, 3$, and not just $(\lambda, \mu) \in S$. Thus, we can multiply both sides of (7.24iii) by $\dot{x}_\mu g^{\mu\mu}$ and sum over $\mu = 0, \ldots, 3$ (in this paragraph we shall write summation signs) to get

$$\sum_{\mu=0}^{3}\left(\dot{x}_\mu{}^2 g^{\mu\mu}g^{\lambda\lambda}\frac{\partial L}{\partial \dot{x}_\lambda} - \dot{x}_\lambda g^{\mu\mu}g^{\mu\mu}\dot{x}_\mu\frac{\partial L}{\partial \dot{x}_\mu}\right) = 0.$$

Applying (7.24ii) to the second term in this equation, we obtain

$$\sum_{\mu=0}^{3}\dot{x}_\mu{}^2 g^{\mu\mu}g^{\lambda\lambda}\frac{\partial L}{\partial \dot{x}_\lambda} = 4L\dot{x}_\lambda, \qquad \lambda = 0, \ldots, 3.$$

Noting that

$$\frac{\partial}{\partial \dot{x}_\lambda}(\log L) = \frac{1}{L}\frac{\partial L}{\partial \dot{x}_\lambda},$$

we can write the last equation as

$$g^{\lambda\lambda}\frac{\partial}{\partial \dot{x}_\lambda}(\log L) = \frac{4\dot{x}_\lambda}{\sum_{\mu=0}^{3}g^{\mu\mu}\dot{x}_\mu{}^2}. \qquad (7.28)$$

By integrating both sides (7.28) with respect to $\dot{x}_\lambda$, we observe that for (7.28) to hold it is sufficient that

$$\log L = 2\log\sum_{\mu=0}^{3}g^{\mu\mu}\dot{x}_\mu{}^2 + \Phi(x),$$

where $\Phi$ is an arbitrary function of the conformal scalar $x = x(t)$. Hence we obtain

$$L = \Phi(x)\left[\sum_{\mu=0}^{3} g^{\mu\mu}\dot{x}_{\mu}^{2}\right]^{2}. \tag{7.29}$$

We shall leave it as an exercise to show that (7.29) satisfies the identity (7.24iv) associated with the inversions. Therefore we have determined a class of Lagrangians which are conformally invariant.

We conclude this section with a result that shows that invariance under dilations alone leads to a homogeneity property of the Lagrangian. First, we recall that a function $u = u(z_1, \ldots, z_n)$ is *positive homogeneous* of degree $p$ if and only if

$$u(tz_1, \ldots, tz_n) = t^p u(z_1, \ldots, z_n), \qquad t > 0.$$

Euler's theorem on positive homogeneous functions can be stated as follows; its proof is simple and can be found in Courant and Hilbert [2], pp. 11–12.

**7.2   Lemma**   If $u(z_1, \ldots, z_n)$ satisfies the first-order partial differential equation

$$z_1 \frac{\partial u}{\partial z_1} + \cdots + z_n \frac{\partial u}{\partial z_n} = pu,$$

then $u(z_1, \ldots, z_n)$ is positive homogeneous of degree $p$.   □

Let us now assume that the Lagrangian for a scalar field depends only upon first derivatives of the field function. If the associated action integral is invariant under dilations, then (7.24ii) implies

$$\dot{x}_\alpha \frac{\partial L}{\partial \dot{x}_\alpha} = 4L,$$

and therefore we have, by Lemma 7.2:

**7.4   Theorem**   If the fundamental integral

$$J(x) = \int_D L(\dot{x}_0, \ldots, \dot{x}_3)\, dt^0 \cdots dt^3$$

is invariant under the dilation

$$\bar{t}^\alpha = t^\alpha + \gamma t^\alpha,$$

then the Lagrangian $L$ must be a positive homogeneous function of degree four.   □

## 7.3   CONFORMAL CONSERVATION LAWS

In Chapter 5 we wrote down conservation laws for scalar fields when the fundamental or action integral was invariant under the Poincaré group, i.e., under translations and proper Lorentz transformations (see Theorem 5.1). To determine conformal conservation laws, we must now consider the dilations and inversions.

From Eq. (5.36) the conservation laws take the form

$$\frac{\partial}{\partial t^\alpha}(H_\beta{}^\alpha \tau_\lambda{}^\beta) = 0, \qquad \lambda = 0, \dots, 3, \tag{7.30}$$

where $H_\beta{}^\alpha$ is the Hamiltonian

$$H_\beta{}^\alpha = -L\delta_\beta{}^\alpha + \dot{x}_\beta \frac{\partial L}{\partial \dot{x}_\alpha}. \tag{7.31}$$

For dilations, the infinitesimal generators $\tau^\alpha$ are given by (7.23iii) (there is just one parameter). Thus, (7.30) becomes

$$\frac{\partial}{\partial t^\alpha}(H_\beta{}^\alpha t^\beta) = 0. \tag{7.32}$$

For inversions, the infinitesimal generators $\tau_\lambda{}^\alpha$ are given by (7.23iv). In this case the conservation laws become

$$\frac{\partial}{\partial t^\alpha}(H_\beta{}^\alpha t^\gamma t_\gamma \delta_\lambda{}^\beta - 2H_\beta{}^\alpha t^\beta t_\lambda) = 0$$

or

$$\frac{\partial}{\partial t^\alpha}(H_\lambda{}^\alpha t_\gamma t^\gamma - 2H_\beta{}^\alpha t^\beta t_\lambda) = 0 \qquad (\lambda = 0, \dots, 3). \tag{7.33}$$

We may summarize the result in the following theorem.

**7.5   Theorem**   If the fundamental integral

$$J(x) = \int_D L(x, \dot{x}_0, \dots, \dot{x}_3)\, dt^0 \cdots dt^3,$$

where $x(t)$ is a scalar field, is invariant under the special conformal transformations (7.21), then fifteen conservation laws for the field are given by (5.47), (5.48), (7.32), and (7.33).   □

**7.2  Example**  If the Lagrange function $L$ is given by (7.29), then it follows that

$$\frac{\partial L}{\partial \dot{x}_\alpha} = 4\Phi(x)\left[\sum_{\mu=0}^{3} g^{\mu\mu}\dot{x}_\mu^{\,2}\right]g^{\alpha\alpha}\dot{x}_\alpha \qquad \text{(no sum on } \alpha).$$

Hence, the Hamiltonian is

$$H_\beta^{\,\alpha} = -\Phi(x)\left(\sum_{\mu=0}^{3} g^{\mu\mu}\dot{x}_\mu^{\,2}\right)\left[\left(\sum_{\mu=0}^{3} g^{\mu\mu}\dot{x}_\mu^{\,2}\right)\delta_\beta^{\,\alpha} - 4g^{\alpha\alpha}\dot{x}_\alpha\dot{x}_\beta\right].$$

Now, in the same manner as in Section 4.4 it is possible to calculate explicitly the conserved quantities; we leave this calculation as an exercise.

## 7.4  CONFORMAL COVARIANCE

We now consider the case of a conformally invariant multiple integral problem

$$J(x) = \int_D L(t, x(t), \partial x(t))\, dt^0 \cdots dt^3, \qquad (7.34)$$

where $x(t) = (x_0(t), \ldots, x_3(t))$ is a covariant vector field. In Chapter 5 we derived the general form of the conservation laws when (7.34) was invariant under translations and proper Lorentz transformations. The form of these laws in the case of dilations and inversions will now be taken up.

We recall that Noether's theorem yields the conservation laws

$$\frac{\partial}{\partial t^\alpha}\left[-H_\beta^{\,\alpha}\tau_\lambda^{\,\beta} + p^{\beta\alpha}\xi_{\beta\lambda}\right] = 0 \qquad (\lambda = 0, \ldots, 3) \qquad (7.35)$$

when the variation integral is invariant under

$$\bar{t}^\alpha = t^\alpha + \varepsilon^\beta\tau_\beta^{\,\alpha}, \qquad \bar{x}_\alpha = x_\alpha + \varepsilon^\beta\xi_{\alpha\beta}, \qquad (7.36)$$

where $\tau_\beta^{\,\alpha}$ and $\xi_{\alpha\beta}$ are the infinitesimal generators, and where

$$H_\beta^{\,\alpha} = -L\delta_\beta^{\,\alpha} + \frac{\partial x_\gamma}{\partial t^\beta}\frac{\partial L}{\partial(\partial x_\gamma/\partial t^\alpha)} \qquad (7.37)$$

and

$$p^{\beta\alpha} = \frac{\partial L}{\partial(\partial x_\beta/\partial t^\alpha)} \qquad (7.38)$$

are the Hamiltonian and momentum complexes, respectively. We also recall from Lemma 5.1 that the $\xi_{\alpha\beta}$ depend upon transformation laws for the field; in the case of a covariant vector field, we have

$$\xi_{\alpha\beta} = -\frac{\partial \tau_\beta{}^\gamma}{\partial t^\alpha} x_\gamma. \tag{7.39}$$

For dilations,

$$\tau^\alpha = t^\alpha,$$

and so from (7.39),

$$\xi_\alpha = -\frac{\partial t^\gamma}{\partial t^\alpha} x_\gamma = -x_\alpha.$$

Thus the conservation law (7.35) takes the form

$$\frac{\partial}{\partial t^\alpha} [H_\beta{}^\alpha t^\beta + p^{\beta\alpha} x_\beta] = 0. \tag{7.40}$$

For inversions, (7.23iv) gives

$$\tau_\lambda{}^\alpha = t^\beta t_\beta \delta_\lambda{}^\alpha - 2t^\alpha t_\lambda.$$

Therefore, (7.39) yields

$$\xi_{\alpha\lambda} = -x_\gamma \frac{\partial}{\partial t^\alpha} (t^\beta t_\beta \delta_\lambda{}^\gamma - 2t^\gamma t_\lambda)$$

$$= 2(x_\alpha t_\lambda - x_\lambda t_\alpha + x_\gamma t^\gamma g_{\alpha\lambda}).$$

From (7.35) the conservation laws under inversions are therefore

$$\frac{\partial}{\partial t^\alpha} [-H_\lambda{}^\alpha t^\gamma t_\gamma + 2H_\beta{}^\alpha t^\beta t_\lambda + 2p^{\beta\alpha}(x_\alpha t_\lambda - x_\lambda t_\alpha + x_\gamma t^\gamma g_{\alpha\lambda})] = 0 \tag{7.41}$$

for $\lambda = 0, \ldots, 3$. We have proved:

**7.6  Theorem**   If the fundamental integral (7.34) is invariant under the fifteen-parameter family of special conformal transformations (7.21), the fifteen conservation laws are given by Eqs. (5.91), (5.92), (7.40), and (7.41).   $\square$

It would be interesting to write down (7.40) and (7.41) for the electromagnetic field. To do so requires knowledge that this field has an action integral which is invariant under the special conformal transformations. But this is a result due to Bessel-Hagen [1] which we now quote.

**7.7 Theorem** The action integral

$$J = -\int_D \tfrac{1}{4} F_{\alpha\beta} F^{\alpha\beta} \, dt^0 \cdots dt^3$$

for the electromagnetic field *in vacuo* is absolutely invariant under the special conformal transformations (7.21). □

This theorem, then, permits us to write down, via Noether's theorem, the conservation laws (7.40) and (7.41) following from invariance under dilations and inversions. We shall treat the dilations, leaving the inversions as an exercise. Recalling Eqs. (5.96) and (5.97) which define the momenta and Hamiltonian, respectively, and substituting into (7.40), we obtain the conservation law

$$\frac{\partial}{\partial t^\alpha} \left[ \frac{t^\alpha}{4} F_{\sigma\rho} F^{\sigma\rho} + F^{\rho\alpha} t_\sigma \frac{\partial A_\rho}{\partial t^\rho} + F^{\sigma\alpha} A_\sigma \right] = 0$$

for dilations, where $A_\rho$, $\rho = 0, \ldots, 3$, is the four-potential.

We have noted in Chapter 5 that invariance under translations and Lorentz transformations lead to conservation of energy, momentum, and angular momentum for the electromagnetic field. The five additional conservation laws obtained by Bessel-Hagen [1] from invariance under the larger special conformal group appear to have no direct physical significance. In fact, Plybon [2] has shown that the four conservation laws obtained from invariance under inversions are functionally dependent upon the remaining eleven (see also Rohrlich [1]).

We now turn to the fundamental invariance identities. Again, we shall derive the identity for the dilations, leaving inversions to the reader. Substituting the generators $\tau^\alpha = t^\alpha$ and $\xi_\alpha = -x_\alpha$ into (4.14) we obtain

$$-\frac{\partial L}{\partial x_\alpha} x_\alpha - 2 \frac{\partial x_\beta}{\partial t^\alpha} \frac{\partial L}{\partial (\partial x_\beta / \partial t^\alpha)} + 4L = 0.$$

If the Lagrangian is independent of the $x_\alpha$, then

$$\frac{\partial x_\beta}{\partial t^\alpha} \frac{\partial L}{\partial (\partial x_\beta / \partial t^\alpha)} = 2L,$$

which implies by Lemma 7.2 that $L$ is homogeneous of degree two in its derivatives. As an example, we note that the electromagnetic field defined by (5.84) satisfies this condition.

Extensions and generalizations of the preceding results can be found in other works. Much of the material in the last few sections is from Logan [4]. The second-order conformally invariant problem is discussed in Logan and Blakeslee [1], and Blakeslee and Logan [1, 2].

## EXERCISES

**7-1**   Show that the expression

$$\cos\theta = \frac{g_{\alpha\beta}(t)a^{\alpha}b^{\beta}}{\sqrt{g_{\alpha\beta}(t)a^{\alpha}a^{\beta}}\sqrt{g_{\alpha\beta}(t)b^{\alpha}b^{\beta}}},$$

which defines the angle $\theta$ between two vectors $a = (a^{\alpha})$ and $b = (b^{\beta})$ at $t$, is independent of coordinate system.

**7-2**   Let $y^{\alpha} = f^{\alpha}(t)$ be a conformal point transformation. Prove that

$$g_{\alpha\beta}(y)\,\frac{\partial y^{\alpha}}{\partial t^{\mu}}\frac{\partial y^{\beta}}{\partial t^{\nu}} = \sigma(t)g_{\mu\nu}(t)$$

if and only if

$$g_{\alpha\beta}(y)\,dy^{\alpha}\,dy^{\beta} = \sigma(t)g_{\alpha\beta}(t)\,dt^{\alpha}\,dt^{\beta}.$$

**7-3**   Prove that the conformal factor for the inversions

$$y^{\alpha} = \frac{t^{\alpha}}{g_{\mu\nu}t^{\mu}t^{\nu}}$$

is given by

$$\sigma(t) = (g_{\mu\nu}t^{\mu}t^{\nu})^{-2}.$$

   [*Hint*: Use Definition 7.1.]

**7-4**   Show that the transformation

$$y^{\alpha} = (t^{\alpha} - \varepsilon^{\alpha}t_{\lambda}t^{\lambda})(1 - 2\varepsilon_{\beta}t^{\beta} + \varepsilon_{\beta}\varepsilon^{\beta}t_{\lambda}t^{\lambda})^{-1}$$

is a product of the two inversions

$$z^{\alpha} = t^{\alpha}(t_{\lambda}t^{\lambda})^{-1}, \qquad w^{\alpha} = (z^{\alpha} - \varepsilon^{\alpha})[(z_{\lambda} - \varepsilon_{\lambda})(z^{\lambda} - \varepsilon^{\lambda})]^{-1} + \varepsilon^{\alpha},$$

and a translation $y^{\alpha} = w^{\alpha} - \varepsilon^{\alpha}$.

**7-5**   Compute the conformal factor for the transformation defined by (7.13).

**7-6**   Show that the Lagrangian

$$L = \Phi(x)\left[\sum_{\mu=0}^{3} g^{\mu\mu}\dot{x}_{\mu}^{2}\right]^{2}$$

satisfies the invariance identity (7.24iv).

**7-7**  Referring to Example 7.2, determine explicitly the conformal conservation laws for the action integral

$$J(x) = \int_D \Phi(x) \left[ \sum_{\mu=0}^{3} g^{\mu\mu} \dot{x}_\mu^2 \right]^2 dt^0 \cdots dt^3,$$

where $x = x(t)$ is a scalar function.

**7-8**  Determine if the action integral

$$J(x) = \int_D \left[ -\tfrac{1}{2} \sum_{\mu=0}^{3} g^{\mu\mu} \dot{x}_\mu^2 - m^2 x^2 \right] dt^0 \cdots dt^3$$

which leads to the Klein–Gordon equation (see Example 5.1) is invariant under the special conformal transformations.

**7-9**  Derive the conservation laws obtained from Noether's theorem for the electromagnetic field in the case that the action integral is invariant under inversions.

**7-10**  Derive invariance identities for the multiple integral problem

$$J = \int_D L(t, x, \partial x)\, dt^0 \cdots dt^3$$

in the case that $J$ is invariant under the special conformal transformations, where $x = (x^0, \ldots, x^3)$ is a covariant vector field (see Logan [4]).

**7-11**  Referring to the last exercise, show that if $L = f(F)$, where $F = F_{\alpha\beta} F^{\alpha\beta}$, $F_{\alpha\beta}$ the electromagnetic field tensor, then $L = kF$, $k$ a constant (see Logan [4]).

# CHAPTER 8

# Parameter-Invariant Problems

## 8.1  INTRODUCTION

Let us consider a variational problem defined by the single integral

$$J(x) = \int_a^b L(t, x, \dot{x}) \, dt, \tag{8.1}$$

where $x \in C_n^2[a, b]$. We make the following definition.

**8.1  Definition**  The fundamental integral (8.1) is *parameter-invariant* if and only if

$$\int_{t_0}^{t_1} L(t, x(t), \dot{x}(t)) \, dt = \int_{\bar{t}_0}^{\bar{t}_1} L\left(\bar{t}, x \circ f^{-1}(\bar{t}), \frac{d}{d\bar{t}} x \circ f^{-1}(\bar{t})\right) d\bar{t} \tag{8.2}$$

for all $[t_0, t_1] \subseteq [a, b]$ and for arbitrary transformations

$$\bar{t} = f(t) \tag{8.3}$$

for which $f \in C^2[a, b]$ and

$$f'(t) > 0 \qquad \text{for} \quad t \in [a, b]. \tag{8.4}$$

150

Definition 8.1 actually requires that the fundamental integral does not change under *arbitrary* transformations of the independent variable $t$. In general, we would not expect a variational problem to be invariant under such transformations. In classical dynamics, for example, the independent variable $t$ is time and it plays an absolute, significant role. However, in problems of geometry such as computing the arclength of a curve, the value of the integral should be independent of any particular parameterization of the curve in question. Therefore, we set aside for study the variational problems for which the arbitrary choice of the independent variables leads to no change in the value of the fundamental integral; these problems are the so-called parameter-invariant problems, and their role in geometry and relativistic mechanics is a significant one. As we shall see, the Lagrangians for such problems are limited severely by the defining condition (8.2).

To offer a few comments on Definition 8.1, we notice that $x \in C_n^2[a, b]$ and $f \in C^2[a, b]$. Condition (8.4) insures the invertibility of $f$, and so it makes sense to consider the composition function $x \circ f^{-1} \in C_n^2[\bar{a}, \bar{b}]$ where $\bar{a} = f(a)$ and $\bar{b} = f(b)$. Definition 8.1 just requires, then, that $J(x) = J(x \circ f^{-1})$ for arbitrary transformations $f$. In many other books, condition (8.2) is usually written

$$\int_{t_0}^{t_1} L(t, x(t), \dot{x}(t)) \, dt = \int_{\bar{t}_0}^{\bar{t}_1} L\left(\bar{t}, \bar{x}(\bar{t}), \frac{d\bar{x}(\bar{t})}{d\bar{t}}\right) d\bar{t}. \tag{8.5}$$

However, when writing (8.5) we must be careful to identify $\bar{x}(\bar{t}) \equiv x \circ f^{-1}(\bar{t})$.

**8.1  Example**  Consider the arclength functional

$$J(x, y) = \int_0^2 \sqrt{\dot{x}^2 + \dot{y}^2} \, dt.$$

Let us subject this to the transformation

$$\bar{t} = f(t) \equiv t^2.$$

Then, clearly, the inverse is given by

$$t = f^{-1}(\bar{t})$$
$$= \sqrt{\bar{t}},$$

and the transformed integral on the right-hand side of (8.2) becomes

$$J(x \circ f^{-1}, y \circ f^{-1}) = \int_0^4 \sqrt{\left(\frac{d}{d\bar{t}} x \circ f^{-1}(\bar{t})\right)^2 + \left(\frac{d}{d\bar{t}} y \circ f^{-1}(\bar{t})\right)^2} \, d\bar{t} \tag{8.6}$$

or, more simply,

$$J(\bar{x}, \bar{y}) = \int_0^4 \sqrt{\left(\frac{d\bar{x}(\bar{t})}{d\bar{t}}\right)^2 + \left(\frac{d\bar{y}(\bar{t})}{d\bar{t}}\right)^2}\ d\bar{t}.$$

By the chain rule for derivatives,

$$\frac{d}{d\bar{t}}\ x \circ f^{-1}(\bar{t})\bigg|_{\bar{t}=f(t)} = \frac{dx \circ f^{-1}(\bar{t})}{dt}\ \frac{dt}{d\bar{t}}\bigg|_{\bar{t}=t^2}$$

$$= \frac{dx(t)}{dt}\ \frac{1}{2t}.$$

A similar expression holds for $y$. Thus, when we make the substitution $\bar{t} = f(t)$ in (8.6) we get

$$J(x \circ f^{-1}, y \circ f^{-1}) = \int_0^2 \sqrt{\left(\frac{\dot{x}(t)}{2t}\right)^2 + \left(\frac{\dot{y}(t)}{2t}\right)^2}\ 2t\ dt$$

$$= \int_0^2 \sqrt{\dot{x}(t)^2 + \dot{y}(t)^2}\ dt$$

$$= J(x, y).$$

Hence, we have shown directly from the definition that the arclength functional is invariant under $\bar{t} = t^2$. It is easy to show that it is invariant under arbitrary transformations $\bar{t} = f(t)$ by repeating the argument above.

**8.2   Example**   Consider the integral

$$J(x) = \int_0^2 \tfrac{1}{2}m\dot{x}^2\ dt$$

and the transformation

$$\bar{t} = f(t)$$
$$= t^2.$$

This integral is not parameter-invariant since

$$J(x \circ f^{-1}) = \int_0^4 \tfrac{1}{2}m\left(\frac{d}{d\bar{t}}\ x \circ f^{-1}(\bar{t})\right)^2\ d\bar{t}$$

$$= \int_0^2 \tfrac{1}{2}m\ \frac{\dot{x}(t)^2}{4t^2}\ 2t\ dt$$

$$= \int_0^2 \frac{m\dot{x}^2}{4t}\ dt$$

$$\neq J(x).$$

**8.3   Example**   In the $xy$-plane suppose we are given the variation problem

$$J(y) = \int_a^b L(x, y(x), y'(x))\, dx \to \min, \tag{8.7}$$

where $y = y(x)$ is the explicit representation of a curve $C$. If we go over to parametric equations

$$x = x(t), \qquad y = y(t) \qquad (t_0 \le t \le t_1)$$

for $C$, then (8.7) can be written

$$J = \int_{t_0}^{t_1} L\left(x(t), y(t), \frac{dy(t)/dt}{dx(t)/dt}\right) \frac{dx}{dt}\, dt$$

$$\equiv \int_{t_0}^{t_1} F(x, y, \dot{x}, \dot{y})\, dt,$$

where $F(x, y, \dot{x}, \dot{y}) \equiv L(x, y, \dot{y}/\dot{x})\dot{x}$. We note that the new Lagrangian $F$ is independent of $t$ (explicitly) and satisfies the condition

$$F(x, y, \lambda\dot{x}, \lambda\dot{y}) = \lambda F(x, y, \dot{x}, \dot{y}),$$

which means it is a positive homogeneous function of degree one in $\dot{x}$ and $\dot{y}$. Consequently, by Euler's Theorem, $F$ must satisfy the first-order partial differential equation

$$\dot{x}F_{\dot{x}} + \dot{y}F_{\dot{y}} = F. \tag{8.8}$$

Example 8.3 shows that an explicit variational problem leads to a problem in parametric form whose Lagrange function is independent of $t$ and satisfies a homogeneity condition. As it turns out, these two conditions are both necessary and sufficient to guarantee parameter-invariance. In addition, the Lagrangians for such problems will be restricted by conditions analogous to (8.8).

## 8.2   SUFFICIENT CONDITIONS FOR PARAMETER-INVARIANCE

We begin with the following theorem which provides conditions under which the integral (8.1) is parameter-invariant.

**8.1   Theorem**   If the Lagrangian $L$ in the fundamental integral (8.1) does not depend explicitly upon $t$, and if it is positive homogeneous of degree one in $\dot{x}$, i.e.,

$$L(x, \lambda\dot{x}) = \lambda L(x, \dot{x}) \qquad (\lambda > 0), \tag{8.9}$$

then the fundamental integral (8.1) is parameter-invariant.

*Proof*  Let $\bar{t} = f(t)$ be an arbitrary transformation of the independent variable satisfying (8.4) and let $a \leq t_0 < t_1 \leq b$. Then

$$\int_{\bar{t}_0}^{\bar{t}_1} L\left(x \circ f^{-1}(\bar{t}), \frac{d}{d\bar{t}} x \circ f^{-1}(\bar{t})\right) d\bar{t} = \int_{t_0}^{t_1} L\left(x(t), \frac{dx(t)}{dt}\frac{dt}{d\bar{t}}\right) \frac{d\bar{t}}{dt} \, dt$$

$$= \int_{t_0}^{t_1} L(x(t), \dot{x}(t)) \, dt,$$

where in the last step we used the homogeneity condition.   $\square$

We can now understand the success and failure of Examples 8.1 and 8.2, respectively. For the arclength functional,

$$L(x, y, \lambda\dot{x}, \lambda\dot{y}) = \sqrt{(\lambda\dot{x})^2 + (\lambda\dot{y})^2}$$
$$= \lambda\sqrt{\dot{x}^2 + \dot{y}^2}$$
$$= \lambda L(x, y, \dot{x}, \dot{y}),$$

so the Lagrangian is positive homogeneous of degree one. However, from Example 8.2 we have

$$L(x, \lambda\dot{x}) = \tfrac{1}{2}m(\lambda\dot{x})^2$$
$$= \tfrac{1}{2}m\lambda^2\dot{x}^2$$
$$= \lambda^2 L(x, \dot{x}),$$

so the given Lagrangian is not homogeneous of degree one, but two.

By differentiating (8.9) with respect to $\lambda$, we get

$$\frac{d}{d\lambda} L(x, \lambda\dot{x}) = L(x, \dot{x}),$$

or, after setting $\lambda = 1$,

$$\frac{\partial L}{\partial \dot{x}^k} \dot{x}^k = L.$$

By Exercise 8.1 this condition is equivalent to the homogeneity condition (8.9); therefore, we have:

**8.2   Theorem**   If the Lagrangian $L$ in the fundamental integral (8.1) satisfies the conditions

$$\frac{\partial L}{\partial t} = 0 \qquad\qquad (8.10)$$

and

$$\dot{x}^k \frac{\partial L}{\partial \dot{x}^k} = L, \tag{8.11}$$

then the fundamental integral (8.1) is parameter-invariant.   □

In the next section we shall show that (8.10) and (8.11) are also necessary.

## 8.3   THE CONDITIONS OF ZERMELO AND WEIERSTRASS

In this section we shall obtain necessary conditions for parameter-invariance. Our approach will be nonstandard in that we apply the results of the general theory of invariant variational problems developed in Chapter 2 rather than give the more common direct arguments. Our calculations are based on the following lemma.

**8.1   Lemma**   If the fundamental integral (8.1) is parameter-invariant, then it is absolutely invariant, in the sense of Definition 2.1, under the one-parameter family of transformations

$$\bar{t} = t + \varepsilon\tau(t), \qquad \bar{x}^k = x^k \tag{8.12}$$

for any $\tau(t)$ satisfying the condition $1 + \varepsilon\tau'(t) \geq 0$.

*Proof*   By hypothesis, (8.2) holds for $f(t) = t + \varepsilon\tau(t)$ since $f'(t) > 0$. But then (8.2) just reduces to the definition of absolute invariance.   □

**8.3   Theorem**   If the fundamental integral (8.1) is parameter-invariant, then the Lagrangian $L$ must satisfy the conditions

$$\frac{\partial L}{\partial t} = 0 \tag{8.13}$$

and

$$\dot{x}^k \frac{\partial L}{\partial \dot{x}^k} = L. \tag{8.14}$$

*Proof*   By Lemma 8.1 we may apply the fundamental invariance theorem, Theorem 2.1. Therefore, with $\tau = \tau(t)$ and $\xi = 0$ the invariance identity becomes

$$\frac{\partial L}{\partial t} \tau + \frac{\partial L}{\partial \dot{x}^k} \left( -\frac{d\tau}{dt} \dot{x}^k \right) + L \frac{d\tau}{dt} = 0$$

or

$$\frac{\partial L}{\partial t} \tau + \left( L - \dot{x}^k \frac{\partial L}{\partial \dot{x}^k} \right) \frac{d\tau}{dt} = 0.$$

Due to the arbitrariness of $\tau$ (choose $\tau \equiv 1$), we have

$$\frac{\partial L}{\partial t} = 0.$$

Choosing $\tau = t$ then gives us

$$L - \dot{x}^k \frac{\partial L}{\partial \dot{x}^k} = 0. \qquad \Box$$

Equations (8.13) and (8.14) are called the *Zermelo conditions*; they were first obtained by Zermelo [1] in 1894 (see Bolza [1], p. 119).

Recalling Example 8.3, we notice that when the explicit problem $\int_a^b L(x, y, y')\, dx$ is replaced by a parametric problem $\int_{t_0}^{t_1} F(x, y, \dot{x}, \dot{y})\, dt$ there results two Euler–Lagrange equations

$$\frac{\partial F}{\partial x} - \frac{d}{dt}\frac{\partial F}{\partial \dot{x}} = 0 \quad \text{and} \quad \frac{\partial F}{\partial y} - \frac{d}{dt}\frac{\partial F}{\partial \dot{y}} = 0 \qquad (8.15)$$

in the place of the single equation

$$\frac{\partial L}{\partial y} - \frac{d}{dx}\frac{\partial L}{\partial y'} = 0$$

for the explicit problem. It is not surprising, therefore, that there should be some connection between the two Euler expressions in (8.15). In general, we have:

**8.4  Theorem**  If the fundamental integral (8.1) is parameter-invariant, then the Euler–Lagrange expressions

$$E_k \equiv \frac{\partial L}{\partial x^k} - \frac{d}{dt}\frac{\partial L}{\partial \dot{x}^k}$$

are connected via the relation

$$\dot{x}^k E_k = 0. \qquad (8.16)$$

*Proof*  Taking the total derivative of (8.14) we obtain

$$\dot{x}^k \frac{d}{dt}\frac{\partial L}{\partial \dot{x}^k} + \frac{\partial L}{\partial \dot{x}^k}\ddot{x}^k = \frac{dL}{dt}$$

or

$$\dot{x}^k \frac{d}{dt}\frac{\partial L}{\partial \dot{x}^k} + \frac{\partial L}{\partial \dot{x}^k}\ddot{x}^k = \frac{\partial L}{\partial t} + \frac{\partial L}{\partial x^k}\dot{x}^k + \frac{\partial L}{\partial \dot{x}^k}\ddot{x}^k.$$

Cancellation and application of (8.13) gives (8.16).  $\Box$

Equation (8.16) is the so-called *Weierstrass representation* for parameter-invariant problems, expressing the fact that the Euler–Lagrange expressions are dependent.

For second-order problems it is possible to determine necessary and sufficient conditions for parameter-invariance in exactly the same manner as for first-order problems. For the second-order variational integral

$$J(x) = \int_a^b L(t, x, \dot{x}, \ddot{x}) \, dt, \quad x \in C_n^4[a, b], \tag{8.17}$$

we define parameter-invariance exactly as in Definition 8.1 except that condition (8.2) is replaced by

$$\int_{t_2}^{t_1} L(t, x, \dot{x}, \ddot{x}) \, dt = \int_{\bar{t}_0}^{\bar{t}_1} L\left(\bar{t}, x \circ f^{-1}(\bar{t}), \frac{d}{d\bar{t}} x \circ f^{-1}(\bar{t}), \frac{d^2}{d\bar{t}^2} x \circ f^{-1}(\bar{t})\right) d\bar{t}. \tag{8.18}$$

Then we have:

**8.5   Theorem**   A necessary condition for the fundamental integral (8.17) to be parameter-invariant is that the Lagrangian satisfy the three conditions

$$\frac{\partial L}{\partial t} = 0, \quad \dot{x}^k \frac{\partial L}{\partial \dot{x}^k} + 2\ddot{x}^k \frac{\partial L}{\partial \ddot{x}^k} = L, \quad \dot{x}^k \frac{\partial L}{\partial \ddot{x}^k} = 0. \tag{8.19}$$

*Proof*   By Lemma 8.1, if (8.17) is parameter-invariant then it is absolutely invariant under the one-parameter family of transformations $\bar{t} = t + \varepsilon\tau(t)$, $\bar{x}^k = x^k$, where $1 + \varepsilon\tau'(t) > 0$. Then, by (6.13) with $\tau = \tau(t)$ and $\xi = 0$,

$$\frac{\partial L}{\partial t}\tau + \left(-\dot{x}^k \frac{\partial L}{\partial \dot{x}^k} - 2\ddot{x}^k \frac{\partial L}{\partial \ddot{x}^k} + L\right)\frac{d\tau}{dt} - \dot{x}^k \frac{\partial L}{\partial \ddot{x}^k}\frac{d^2\tau}{dt^2} = 0.$$

Since $\tau$ is arbitrary, by successively choosing $\tau$ equal to 1, $t$, and $t^2$ we obtain (8.19).   □

The three conditions (8.19) are the Zermelo conditions for the second-order, parameter-invariant problem. These conditions may easily be shown to be sufficient as well. Also, as in the first-order case, the Euler–Lagrange expressions $E_k^{(2)}$ are not independent, but rather

$$\dot{x}^k E_k^{(2)} = 0, \tag{8.20}$$

which is the Weierstrass condition for the second-order problem.

For multiple integral problems, parameter-invariance is discussed in Rund [1, 2, 3], Lovelock and Rund [1], and Hilbert [1, 2].

## 8.4  THE SECOND NOETHER THEOREM

In this section we shall present a special case of the second Noether theorem on invariant variational problems. Let us again consider the fundamental integral (8.1). The invariance transformations that we shall consider are transformations that depend upon an arbitrary function $p(t)$ and its derivatives up to some fixed order $r$. In particular, let

$$\bar{t} = t, \qquad \bar{x}^k = x^k + T^k(p) \tag{8.21}$$

be a family of transformations on $(t, x)$ space depending upon an arbitrary function $p \in C^{r+2}[a, b]$, where the $T^k$, $k = 1, \ldots, n$, are linear differential operators

$$T^k \equiv a_0{}^k(t) + a_1{}^k(t)\frac{d}{dt} + \cdots + a_r{}^k(t)\frac{d^r}{dt^r},$$

with the coefficients $a_i{}^k \in C^2[a, b]$. Then, each curve $x \in C_n{}^2[a, b]$ gets mapped via the transformation to another curve $\bar{x} = \bar{x}(t)$ in $C_n{}^2[a, b]$ defined by

$$\bar{x}^k(t) = x^k(t) + a_0{}^k(t)p(t) + a_1{}^k(t)p'(t) + \cdots + a_r{}^k(t)p^{(r)}(t)$$
$$= x^k(t) + T^k(p(t)).$$

Of course, $\bar{x}(t)$ depends on the choice of the arbitrary function $p(t)$. Now we define invariance.

**8.2  Definition**  The fundamental integral (8.1) is invariant under transformations (8.21) if and only if for all $x \in C_n{}^2[a, b]$ we have

$$\int_a^b L(t, \bar{x}(t), \dot{\bar{x}}(t))\, dt = \int_a^b L(t, x(t), \dot{x}(t))\, dt + o(p),$$

where $o(p)$ denotes higher order terms in $T^k(p)$ and $dT^k(p)/dt$.

We now proceed with a formal calculation which will lead to an invariance identity which relates the Euler–Lagrange expressions and their derivatives up to order $r$. By expanding $L(t, \bar{x}, \dot{\bar{x}})$ in a Taylor series about $(t, x, \dot{x})$ we obtain

$$L(t, \bar{x}, \dot{\bar{x}}) = L(t, x, \dot{x}) + \frac{\partial L}{\partial x^k}T^k(p) + \frac{\partial L}{\partial \dot{x}^k}\frac{d}{dt}T^k(p) + o(p).$$

Integrating from $a$ to $b$ gives

$$\int_a^b L(t, \bar{x}, \dot{\bar{x}})\, dt = \int_a^b L(t, x, \dot{x})\, dt + \int_a^b \left(\frac{\partial L}{\partial x^k}T^k(p) + \frac{\partial L}{\partial \dot{x}^k}\frac{d}{dt}T^k(p)\right) dt + o(p).$$

If the fundamental integral (8.1) is invariant under (8.21), then Definition 8.2 implies

$$\int_a^b \left( \frac{\partial L}{\partial x^k} T^k(p) + \frac{\partial L}{\partial \dot{x}^k} \frac{d}{dt} T^k(p) \right) dt = 0.$$

The second term in the integrand may be integrated by parts to give

$$\int_a^b E_k T^k(p) \, dt + \left[ \frac{\partial L}{\partial \dot{x}^k} T^k(p) \right]_a^b = 0.$$

Since $p(t)$ is arbitrary, we may choose $p(t)$ such that $p(a) = p'(a) = \cdots = p^{(r)}(a) = 0$ and $p(b) = p'(b) = \cdots = p'(b) = \cdots = p^{(r)}(b) = 0$. Therefore the boundary term in the last equation vanishes and we are left with

$$\int_a^b E_k T^k(p) \, dt = 0. \tag{8.22}$$

Now, we define the *adjoint operator* $\tilde{T}$ of a differential operator $T$ by the Lagrange identity

$$\int_a^b q T(p) \, dt = \int_a^b p \tilde{T}(q) \, dt + [\cdot]_a^b, \tag{8.23}$$

where $q \in C^r[a, b]$ and $[\cdot]_a^b$ represent boundary terms; this can be obtained by the repeated integration by parts of the left-hand side of (8.23). Using (8.23), we may write (8.22) as

$$\int_a^b \tilde{T}^k(E_k) p \, dt + [\cdot]_a^b = 0.$$

Again appealing to the arbitrariness of $p(t)$ we can force the boundary term to vanish, after which we can apply the fundamental lemma of the calculus of variations to conclude that

$$\tilde{T}^k(E_k) = 0. \tag{8.24}$$

Equation (8.24) is the *Noether identity*; we summarize our formal calculation in the following:

**8.6    Theorem** If the fundamental integral (8.1) is invariant under the transformation (8.21) depending upon an arbitrary function $p(t)$, then

$$\tilde{T}^k(E_k) = 0,$$

where the $E_k$ are the $n$ Euler–Lagrange expressions and $\tilde{T}^k$ is the adjoint of $T^k$ in (8.21).    □

We can clearly extend this result to multiple integral problems. We do so in the case $T^k$ is a first-order differential operator.

**8.7  Theorem**  If the multiple integral

$$J(x) = \int_D L(t, x(t), \partial x(t))\, dt^1 \cdots dt^m \qquad (x \in C_n^2(D))$$

is invariant under the transformation

$$\bar{t}^\alpha = t^\alpha, \qquad \bar{x}^k = x^k + T^k(p) \qquad (p \in C^3(D)), \tag{8.25}$$

where

$$T^k = a^k(t) + a_\alpha{}^k(t)\frac{\partial}{\partial t^\alpha} \qquad (a^k, a_\alpha{}^k \in C^2(D)),$$

then

$$\tilde{T}^k(E_k) = 0. \qquad \square \tag{8.26}$$

**8.4  Example**  From Chapter 5, the electromagnetic field in empty space was described in terms of the four-potential $A_0, \ldots, A_3$ in which Maxwell's equations took the form

$$E_k \equiv \frac{\partial^2 A_k}{\partial t^{0\,2}} - \nabla^2 A_k = 0 \qquad (k = 0, \ldots, 3)$$

(see (5.73) and (5.74)). We noted that a gauge transformation (see (5.70))

$$\bar{A}_k = A_k + \frac{\partial p}{\partial t^k} \qquad (p \text{ arbitrary}) \tag{8.27}$$

does not affect the definition of the field, and therefore the action integral (5.81) is *gauge-invariant*, or invariant under (8.27). Therefore, Theorem 8.7 can be applied with

$$T^k \equiv \frac{\partial}{\partial t^k}, \qquad x^k \equiv A_k.$$

It is clear that $\tilde{T}^k = -\partial/\partial t^k$, and therefore Eq. (8.26) gives the identity)

$$-\frac{\partial}{\partial t^k}\left(\frac{\partial^2 A_k}{\partial t^{0\,2}} - \nabla^2 A_k\right) = 0.$$

We conclude this chapter with some additional, but brief, remarks on the second Noether theorem. As we have already commented, Theorems 8.6

and 8.7 are restricted versions of the theorem. The general version permits transformations of the form

$$\bar{t} = \phi(t, x, p(t), p'(t), \ldots, p^{(r)}(t)),$$
$$\bar{x}^k = \psi^k(t, x, p(t), p'(t), \ldots, p^{(r)}(t)),$$

$$(8.28)$$

where $p \in C^{r+2}[a, b]$ is an arbitrary function, and where $\bar{t} = t$ and $\bar{x}^k = x^k$ when $p(t) \equiv p'(t) \equiv \cdots \equiv p^{(r)}(t) \equiv 0$. By expanding (8.28) in a Taylor series, there results the infinitesimal transformation

$$\bar{t} = t + U(p), \qquad \bar{x}^k = x^k + T^k(p),$$

where $U$ and $T^k$ are linear differential operators of order $r$. Invariance of (8.1) under (8.28) then implies the existence of an identity of the form

$$\tilde{T}^k(E_k) - \tilde{U}(\dot{x}^k E_k) = 0. \tag{8.29}$$

In the case $p$ is a vector function $p = (p_1, \ldots, p_s)$, then $s$ identities of the form (8.29) result. Equation (8.29) is the general Noether identity for problems invariant under an infinite continuous group. The proof of (8.29) is generally carried out following Noether's original argument based on the fundamental variational formula (see Exercise 8.7). This account can be found in Noether [1] and Funk [1]. An argument similar to that in Chapter 2 can be found in Logan [3]. For applications we mention Hilbert [1, 2] and Drobot and Rybarski [1]. A modern version of the second Noether theorem in the language of modern differential geometry has been given by Komorowski [2].

Finally, it is interesting to note that the Weierstrass conditions for parameter-invariance follow easily from (8.29). For, if the integral (8.1) is parameter-invariant, then it is invariant under transformations of the form

$$\bar{t} = t + p(t), \qquad \bar{x}^k = x^k.$$

Thus $U \equiv I$, the identity operator, and $T^k \equiv 0$. Therefore, $\tilde{U} = I$ and $\tilde{T}^k \equiv 0$ and it follows from (8.29) that

$$\dot{x}^k E_k = 0,$$

which is the Weierstrass condition.

## EXERCISES

**8-1** Prove Euler's theorem: Let $u(x^1, \ldots, x^n)$ be of class $C^1$. Then $u$ is positive homogeneous of degree $k$ if and only if

$$x^1 u_{x^1} + \cdots + x^n u_{x^n} = ku.$$

**8-2**  Determine which of the following variational integrals are parameter-invariant:

(a)  $J(x, y) = \int_a^b \sqrt{1 + \dot{x}^2 + \dot{y}^2}\ dt$;

(b)  $J(x) = \int_a^b (\ddot{x}^2 + 2x\dot{x})\ dt$;

(c)  $J(x) = \int_a^b g_{\alpha\beta}(x)\dot{x}^\alpha \dot{x}^\beta\ dt$;

(d)  $J(x, y) = \int_a^b \eta(x, y)\sqrt{\dot{x}^2 + \dot{y}^2}\ dt$.

**8-3**  Derive Eq. (8.20).

**8-4**  Show that if the fundamental integral

$$J(x) = \int_a^b L(x, \dot{x})\ dt$$

is parameter-invariant, then $L$ satisfies the condition

$$L(x, \lambda\dot{x}) = \lambda^2 L(x, \dot{x}).$$

**8-5**  Prove the converse of Theorem 8.5: If the Lagrangian satisfies the conditions (8.19), then

$$J(x) = \int_a^b L(t, x, \dot{x}, \ddot{x})\ dt$$

is parameter-invariant.

**8-6**  Let $F(x, y, \dot{x}, \dot{y}) \in C^3$ be a positive homogeneous function of degree one.

  (a)  Show that $F_{\dot{x}}$ and $F_{\dot{y}}$ are positive homogeneous of degree zero.
  (b)  Show that $F_{\dot{x}\dot{x}}, F_{\dot{y}\dot{y}}$, and $F_{\dot{x}\dot{y}}$ are positive homogeneous of degree $-1$.

**8-7**  The fundamental variational formula for the first variation of the integral (8.1), when both the independent and dependent variables $t$ and $x^k$ are subjected to small variations $\Delta t$ and $\Delta x^k$, is given by (see Gelfand and Fomin [1] or Rund [1])

$$\Delta J(\Delta t, \Delta x^k) = \int_a^b E_k(\Delta x^k - \dot{x}^k\ \Delta t)\ dt + \left[\left(L - \dot{x}^k \frac{\partial L}{\partial \dot{x}^k}\right)\Delta t + \frac{\partial L}{\partial \dot{x}^k} \Delta x^k\right]_a^b.$$

Defining the invariance of (8.1) under (8.28) by the condition

$$\Delta J(U(p), T^k(p)) = 0,$$

show that

$$\tilde{T}^k(E_k) - \tilde{U}(\dot{x}^k E_k) = 0,$$

thus giving a formal derivation of the second Noether theorem.

**8-8** Give a proof of the first Noether theorem based on the fundamental variational formula given in Exercise 8.7.

**8-9** Prove that a necessary condition for the fundamental integral (8.17) to be parameter-invariant is that

$$\det\left(\frac{\partial^2 L}{\partial \ddot{x}^k \, \partial \ddot{x}^j}\right) = 0.$$

# References

AKHEIZER, N. I.
1. "The Calculus of Variations." Ginn (Blaisdell), Boston, 1962.
ANDERSON, D.
1. Noether's Theorem in Generalized Mechanics, *J. Phys. A* **6**, 299–305 (1973).
BATEMAN, H.
1. The Transformation of the Electrodynamical Equations, *Proc. London Math. Soc.* **7**, 223–264 (1910).
BARUT, A. O.
1. "Electrodynamics and Classical Theory of Fields and Particles." Macmillan, New York, 1964.
BESSEL-HAGEN, E.
1. Über die Erhaltungssätze der Elektrodynamik, *Math. Ann.* **84**, 258–276 (1921).
BISHOP, R. L., and GOLDBERG, S. I.
1. "Tensor Analysis on Manifolds." Macmillan, New York, 1968.
BLAKER, J. W., and TAVEL, M. A.
1. The Application of Noether's Theorem to Optical Systems, *Amer. J. Phys.* **42**, 857–861 (1974).
BLAKESLEE, J. S.
1. Invariance Identities for Higher-Order Variational Problems, Dissertation, Kansas State University, Manhattan, (1975).
BLAKESLEE, J. S., and LOGAN, J. D.
1. Conformal Conservation Laws for Second-Order Scalar Fields, *Nuovo Cimento* **34B**, 319–324 (1976).
2. Conformal Identities and Conservation Laws for Second-Order Variational Problems Depending on a Covariant Vector Field, *J. Phys. A* (in press).
BLISS, G. A.
1. "Calculus of Variations" (Carus Math. Mono. **1**). Open Court Publ., LaSalle, Illinois, 1925.

BOLZA, O.
1. "Vorlesungen über Variationrechnung." Chelsea Publ., Bronx, New York (Reprint of the 1909 version).
2. "Lectures on the Calculus of Variations." Dover, New York, 1961.

BOGOLUIBOV, N. N., and SHIRKOV, D. V.
1. "Introduction to the Theory of Quantized Fields." Wiley (Interscience), New York, 1959.

COURANT, R., and HILBERT, D.
1. "Methods of Mathematical Physics," Vol. 1. Wiley (Interscience), New York, 1953.
2. "Methods of Mathematical Physics," Vol. 2, Partial Differential Equations. Wiley (Interscience), New York, 1962.

CUNNINGHAM, E.
1. The Principle of Relativity in Electrodynamics and an Extension Thereof, *Proc. London Math. Soc., Sec. Ser.* **8**, 77–98 (1910).

DAVIS, H. T.
1. "Introduction to Nonlinear Differential and Integral Equations." Dover, New York, 1962.

DAVIS, W. R.
1. "Classical Fields, Particles, and the Theory of Relativity." Gordon & Breach, New York, 1970.

DROBOT, S., and RYBARKSI, A.
1. A Variational Principle in Hydromechanics, *Arch. Rat. Mech. Anal.* **2** (5), 393–410 (1958–1959).

EISENHART, L. P.
1. "Continuous Groups of Transformations." Princeton Univ. Press, Princeton, New Jersey, 1933.

EDELEN, D. G. B.
1. "Nonlocal Variations and Local Invariance of Fields." Amer. Elsevier, New York, 1969.

ELSGOLC, L. E.
1. "Calculus of Variations." Addison-Wesley, Reading, Massachusetts, 1962.

EPHESER, H.
1. "Vorlesung über Variationsrechnung." Vandenhoeck and Ruprecht, Göttingen, 1973.

FULTON, T., ROHRLICH, F., and WITTEN, L.
1. Conformal Invariance in Physics, *Rev. Mod. Phys.* **34** (3), 442–457 (1962).

FUNK, P.
1. "Variationsrechnung und ihre Anwendung in Physik und Technik." Springer-Verlag, Berlin, 1962.

GARCÍA, P. L.
1. Geometriá Simplectica en la Theoría Clásica de Campos, *Collect. Math.* **19**, 73–134 (1968).

GELFAND, I. M., and FOMIN, S. V.
1. "Calculus of Variations." Prentice-Hall, Englewood Cliffs, New Jersey, 1963.

GOLDSTEIN, H.
1. "Classical Mechanics." Addison-Wesley, Reading, Massachusetts, 1950.

GRÄSSER, H. S. P.
1. A Monograph on the General Theory of Second-Order Parameter Invariant Problems in the Calculus of Variations, *Math. Commun. Univ. South Africa* **M2**, Pretoria (1967).

HAANTJES, J.
1. Conformal Representations of an $n$-dimensional Euclidean Space with a Non-Definite Fundamental Form on Itself, *Nederl. Akad. Wetensch. Amsterdam Proc.* **40**, 700–705 (1937).

HILBERT, D.
1. Die Grundlagen der Physik, *Nachr. Ges. Wiss. Göttingen, Math.-phys. Kl.* 395–407 (1915); 53–76 (1916).
2. Grundlagen der Physik, *Math. Ann.* **92**, 258–289 (1924).

HILL, E. L.
1. Hamilton's Principle and the Conservation Theorems of Mathematical Physics, *Rev. Mod. Phys.* **23** (3), 253–260 (1951).

HSU, C.-J. (T. OHYAMA)
1. A Remark on the Extension of Liouville's Theorem to a Euclidean Space of Signature (+, +, −), *Tôhoku Math. J.* **49** (2), 139–144 (1943).

JACKSON, J. D.
1. "Classical Electrodynamics" (Second Ed.). Wiley, New York, 1975.

KILLING, W.
1. Über die Grundlagen der Geometrie, *J. Reine Angew. Math.* **109**, 121–186 (1892).

KLEIN, F.
1. Über die Differentialgesetze für die Erhaltung von Impuls und Energie in der Einsteinschen Gravitationstheorie, *Nachr. Akad. Wiss. Göttingen, Math.-Phys. Kl. II* **1918**, 171–189 (1918).

KLÖTZLER, R.
1. "Mehrdimensionale Variationsrechnung." Birkhaeuser, Basel, 1970.

KOMOROWSKI, J.
1. A Modern Version of E. Noether's Theorem in the Calculus of Variations, I, *Studia Math.* **29**, 261–273 (1968).
2. A Modern Version of E. Noether's Theorem in the Calculus of Variations, II, *Studia Math.* **32**, 181–190 (1968).

LANCZOS, C.
1. "The Variational Principles of Mechanics." Univ. of Toronto Press, Toronto, 1949.

LANDAU, L. D., and LIFSCHITZ, E. M.
1. "The Classical Theory of Fields" (Fourth Ed.). Pergamon Press, New York, 1975.

LIE, S.
1. "Vorlesungen über differentialgleichungen mit bekannten infinitesimalen Transformationen." Teubner, Leipzig, 1912.

LOGAN, J. D.
1. Invariance and the $n$-Body Problem, *J. Math. Anal. Appl.* **43** (1), 191–197 (1973).
2. First Integrals in the Discrete Variational Calculus, *Aequationes Math.* **9** (2/3), 210–220 (1973).
3. On Variational Problems which Admit an Infinite Continuous Group, *Yokohama Math. J.* **22** (1-2), 31–42 (1974).
4. Conformal Invariance of Multiple Integrals in the Calculus of Variations, *J. Math. Anal. Appl.* **48** (2), 618–631 (1974).
5. On Some Invariance Identities of H. Rund, *Utilitas Math.* **7**, 281–286 (1975).

LOGAN, J. D., and BLAKESLEE, J. S.
1. An Invariance Theory for Second-Order Variational Problems, *J. Mathematical Phys.* **16** (7), 1374–1379 (1975).

LOVELOCK, D. and RUND, H.
1. "Tensors, Differential Forms, and Variational Principles." Wiley (Interscience), New York, 1975.

McFARLAN, L. H.
1. Problems of the Calculus of Variations in Several Dependent Variables and their Derivatives of Various Orders, *Tôhoku Math. J.* **33**, *Ser. 1*, 204–218 (1931).

MARION, J. B.
1. "Classical Dynamics of Particles and Systems" (Second Ed.). Academic Press, New York, 1970.

MARSDEN, J.
1. "Applications of Global Analysis in Mathematical Physics." Publish or Perish, Boston, 1974.

MIURA, R. M.
1. The Korteweg–deVries Equation: A Survey of Results, *SIAM Rev.* **18** (3), 412–459 (1976).

MORREY, C. B.
1. "Multiple Integrals in the Calculus of Variations." Springer-Verlag, Berlin and New York, 1966.

MISNER, C. W., THORNE, K. S., and WHEELER, J. A.
1. "Gravitation." Freeman, San Francisco, 1973.

NAÏMARK, M. A.
1. "Linear Representations of the Lorentz Group." Macmillan, New York, 1964.

NOETHER, E.
1. Invariante Variationsprobleme, *Nachr. Akad. Wiss. Gottingen, Math.-Phys. Kl. II* **1918**, 235–257 (1918).
2. Invariant Variation Problems, *Transport Theory of Statis. Physics* **1**, 186–207 (1971) (Translation by M. A. Tavel of the original article).

PARS, L. A.
1. "A Treatise on Analytical Dynamics." Wiley, New York, 1965.

PLYBON, B.
1. Conservation Laws for Electromagnetic Fields, Dissertation, Ohio State University, Columbus (1968).

ROHRLICH, F.
1. "Classical Charged Particles." Addison-Wesley, Reading, Massachusetts, 1965.

RUND, H.
1. "The Hamilton–Jacobi Theory in the Calculus of Variations." Van Nostrand–Reinhold, Princeton, New Jersey, 1966.
2. Variational Problems Involving Combined Tensor Fields, *Abh. Math. Sem. Univ. Hamburg* **29**, 243–262 (1966).
3. Invariant Theory of Variational Problems for Geometric Objects, *Tensor* **18**, 239–258 (1967).
4. A Direct Approach to Noether's Theorem in the Calculus of Variations, *Utilitas Math.* **2**, 205–214 (1972).

SAGAN, H.
1. "Introduction to the Calculus of Variations." McGraw-Hill, New York, 1969.

SPIVAK, M.
1. "Differential Geometry." Publish or Perish, Boston; Vol. I, 1970; Vol. II, 1970.

TRAUTMAN, A.
1. Noether's Equations and Conservation Laws, *Commun. Math. Phys.* **6**, 248–261 (1967).

WEINSTOCK, R.
1. "Calculus of Variations." McGraw-Hill, New York, 1952; reprinted by Dover, New York, 1975.

WHITHAM, G. B.
1. "Linear and Nonlinear Waves." Wiley (Interscience), New York, 1974.

WHITTAKER, E. T.
1. "A Treatise on the Analytical Dynamics of Particles and Rigid Bodies" (Fourth Ed.). Cambridge Univ. Press, Cambridge, 1937; reprinted by Dover, New York, 1944.

ZERMELO, E.
1. Untersuchungen zur Variationsrechnung, Dissertation, Berlin (1894).

# Index

A
B  7
C  8
D  9
E  0
F  1
G  2
H  3
I  4
J  5